North York Moors and Yorkshire Wolds

LANDSCAPE AND GEOLOGY

North York Moors and Yorkshire Wolds

LANDSCAPE AND GEOLOGY

TONY WALTHAM

THE CROWOOD PRESS

Contents

CHAPTER 1

Moors and Wolds of Yorkshire

Diversity is a simple delight in the landscapes of eastern Yorkshire, and it is all down to the geology. Sandstones define the North York Moors, and their extension into the Cleveland Hills, with their grand panoramas across the empty and lonely wilderness of heather moorland. Chalk creates a totally different terrain in the Yorkshire Wolds, with its rolling hills all clad in farmland except for some of their deeper and steeper valleys. In between the Moors and Wolds, limestone characterises landscapes in the Tabular Hills and Howardian Hills, with farmland supporting a suite of lovely stone-built villages. Those hills overlook the Vale of Pickering that owes its remarkably flat floor to sediments deposited in a bygone lake. Then farther south, inside the great curve of the Wolds, Holderness is an extensive lowland formed on a veneer of glacial till left over from the recent Ice Ages.

Diversity, even contrast, appears again in the long Yorkshire coastline. Sweeping bays and headlands fringe the North York Moors; they vary in detail due to erosion picking out their individual combinations of strong sandstones and soft clays. Then the eroded end of the Yorkshire Wolds provides a complete contrast with its towering cliffs of white chalk. And different again, the coast of Holderness has a long line of low cliffs that are collapsing and retreating in dramatic style because they are formed of weak and crumbly glacial till.

The full story of the landscapes of eastern Yorkshire reaches back some 200 million years, when the first sedimentary rocks were formed to constitute the fabric of the terrain. Anything

The view south down Farndale, with its farmland floor between heather moors on the sandstone ridges, with the road along Blakey Ridge visible on the left.

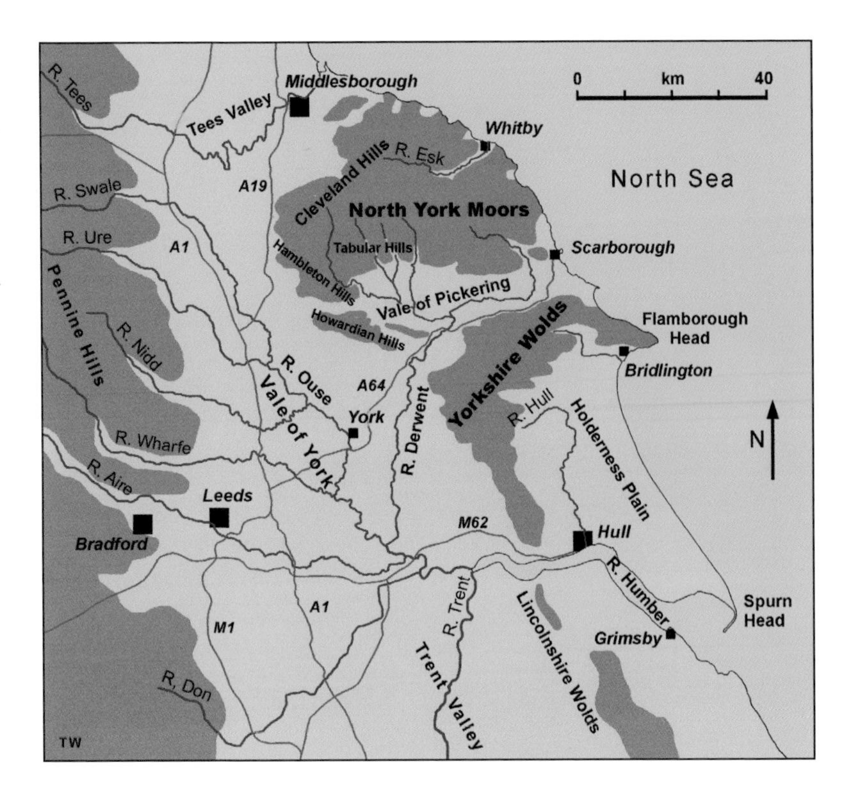

Main features of eastern Yorkshire, with the North York Moors and the Yorkshire Wolds forming the large blocks of higher ground that lie between the broad sweep of the Vale of York and the North Sea coastline.

and everything older than that still exists, but is buried so far down to be almost irrelevant. The northern half of our region, the Moors and adjacent hills, are made of Jurassic rocks, which originated as layers of sand and mud deposited in shallow seas; and they differ from Jurassic rocks elsewhere in England because they accumulated in their own sea almost isolated north of a contemporary strip of land roughly where Market Weighton now stands. It was then all change in Cretaceous times when shell debris rained down on the floor of a tropical sea that spread over the entire region; this eventually became the chalk that makes up the Wolds in its southern half.

Rocks are old but landscapes are younger. The hills and valleys, the texture of the landscapes that we see today, have evolved largely within the last few million years. Pennine rivers played their part in the early stages, and the largest landforms are still dictated by the underlying geology. But much of the detail was added within the last 30,000 years, notably during a brief interlude that saw glaciers surround the hills and re-direct the all-powerful river erosion.

The classic view of Whitby seen from the West Cliff, with the Abbey ruins atop the East Cliff on the far side of the harbour. The jawbones of a bowhead whale form the arch; whaling from Whitby ended in 1833, and these replacement bones came from Inuit hunters in Alaska in 2003.

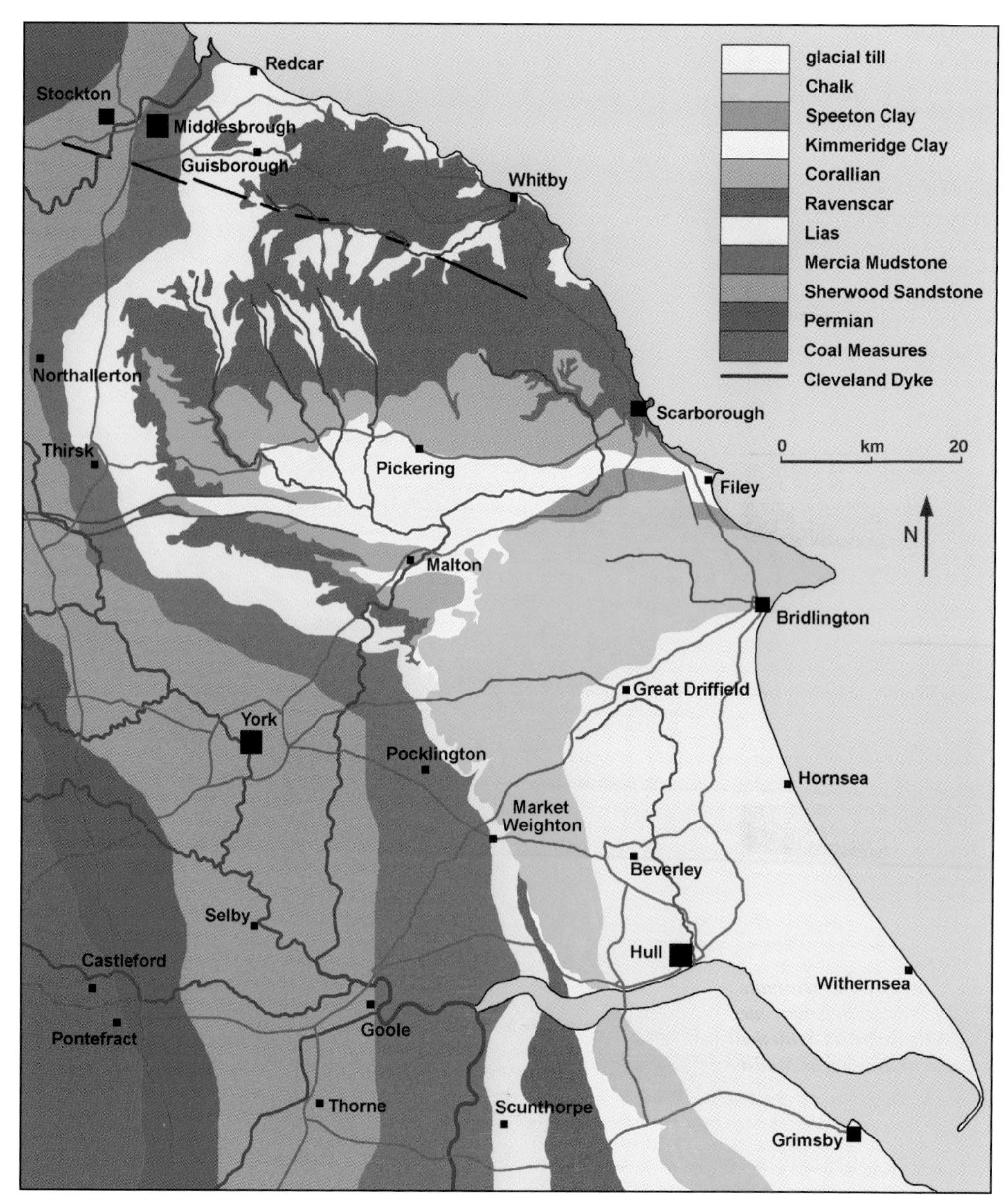

Main features of the geology of the Moors and Wolds, and adjacent areas, in East Yorkshire. Glacial till is only shown in its great spread across Holderness, with its continuation south of the Humber.

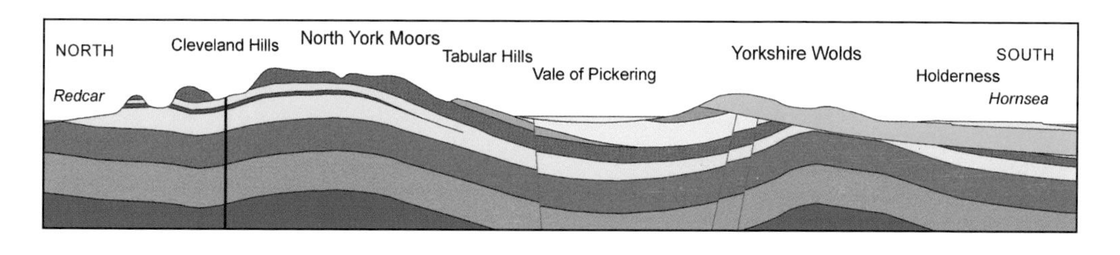

ages (Ma)	main units	upland / lowland
Quaternary	glacial till and lacustrine	Holderness and Vale of Pickering
66		
Cretaceous	Chalk	Yorkshire Wolds
145	Speeton Clay	Vale of Pickering
Upper Jurassic	Kimmeridge Clay	
	Corallian grits and oolites	Hambleton Hills and Tabular Hills
161	Oxford Clay	
Middle Jurassic 176	Ravenscar sandstones	North York Moors Cleveland Hills Howardian Hills
Upper Lias	Whitby Mudstone	northern fringes of the Moors
Middle Lias	Cleveland Ironstone	
Lower Lias 201	Redcar Mudstone	Robin Hood's Bay and eastern edge of Vale of York
Triassic	Mercia Mudstone	Vale of York

Stratigraphy of the main rock units of the Moors and Wolds, along with their main roles in the landscapes. The Lias forms the Lower Jurassic. Units are shown in rough proportion with their local thicknesses: from the base of the Redcar Mudstone to the top of the Chalk, the sedimentary rock sequence totals around 1600 metres thick. Ages of the bases of the main units (and the top of the chalk) are shown in Ma (millions of years before the present). Quaternary sediments are discontinuous and rarely more than 40 metres thick; their age ranges from 2.58 Ma until today.

OPPOSITE ***An outline profile of the geology of the Moors and Wolds, generalised and simplified along a line from Redcar to Hornsea. Key to the rock units is as in the map above, with the addition of the ironstones of the Cleveland Hill, marked in purple, and the lacustrine sediments of the Vale of Pickering, marked in yellow. This profile has a greatly exaggerated vertical scale: it is nearly 100 km long and extends to about 1000 metres below sea level.***

So much for the natural part of the landscape story, before mankind stepped in to make his own mark. Yet again, this was dictated by the geology, with a long history of mining around the northern parts of the Moors. The world's first major chemical industry started in the 1600s with alum produced from shales that were dug out on a huge scale, particularly along the coast of the Moors. Then the 1800s saw ironstone mining in the Cleveland Hills that briefly dominated world production. Both those mining industries left their mark on the landscape, unlike the modern potash mining that has Britain's deepest mines almost unseen beneath the Moors.

Evolution of the landscape never ceases while rainfall, streams and rivers continue their persistent erosion of the ground surface. But geological processes are slow, and change is rarely noticeable within a single person's lifetime. The exception is on some coastlines, and the Holderness coast is among Britain's most dramatic, where land and villages are being lost to the relentless sea as it eats into low cliffs of ground too weak to offer any serious resistance.

Those lost villages of Holderness are one facet of mankind's interaction with the natural world. Rather more successful has been the interaction that has developed the farmland defining most of today's landscapes. The exception to that is the swathe of heather moors that distinguish the core of the North York Moors, designated as a National Park since 1952. The big growth industry of modern times has been tourism, where the coast, both within and beyond the national park, is always a prime attraction. Scarborough became one of England's great Victorian resort towns. It still stands proud within a string of delightful seaside villages and towns, but the popularity of the coastal footpaths is perhaps an indicator of a trend towards the natural world as the big visitor appeal. It is good that eastern Yorkshire has so much to offer in the landscapes of the Moors and Wolds.

CHAPTER 2

Jurassic Rocks of the Moors

A little more than 200 million years ago the world was slowly warming and sea levels were rising. Britain was then at a latitude of about 35 degrees, the same as northern Morocco today, and was just part of a great desert far from any ocean while the supercontinent of Pangaea was still intact. Uplands of hard old rocks, essentially Scotland, the Pennines and Wales, had survived millions of years of slow desert erosion, but most of England was a low desert plain where clays and silts accumulated to form the Mercia Mudstone that now underlies the Vale of York. However, the whole region was then subjected to rifting and subsidence when the Atlantic Ocean started to open up not far to the west.

The combination of rising sea level and sinking land saw the marine invasion that brought shallow Jurassic seas across eastern Yorkshire and nearly all of south-eastern England. Marine sedimentation set in to form the Jurassic and Cretaceous rock sequence. Sea depths varied and coastlines shifted, but sedimentary rocks were formed in huge quantities, except at Market Weighton, where an island soon emerged from the sea and separated the Cleveland Basin from the rest of England to the south. This is why the Jurassic rocks of the North York Moors are conspicuously different from the well-known oolitic limestones of the Midlands, Cotswolds and southern Jurassic Coast.

Land Barrier at Market Weighton

A significant feature of the Jurassic terrains is known as the Market Weighton Block because it can be recognised by the thinning and absence of nearly all the Jurassic beds along the foot of the Wolds escarpment for about 20 km north of the market town. That is along the outcrop, and any extension of the Market Weighton land to the west can only be surmised, as all has been removed by erosion; it remains uncertain whether or not there was a seaway between the Pennine landmass and the Market Weighton land. However, it is known from boreholes that the Jurassic land occupied much of the area now beneath the Wolds; it also extended for around 30 km east of today's coastline.

Staithes Sandstone, of the Middle Lias, forming the cliffs around the village of Staithes.

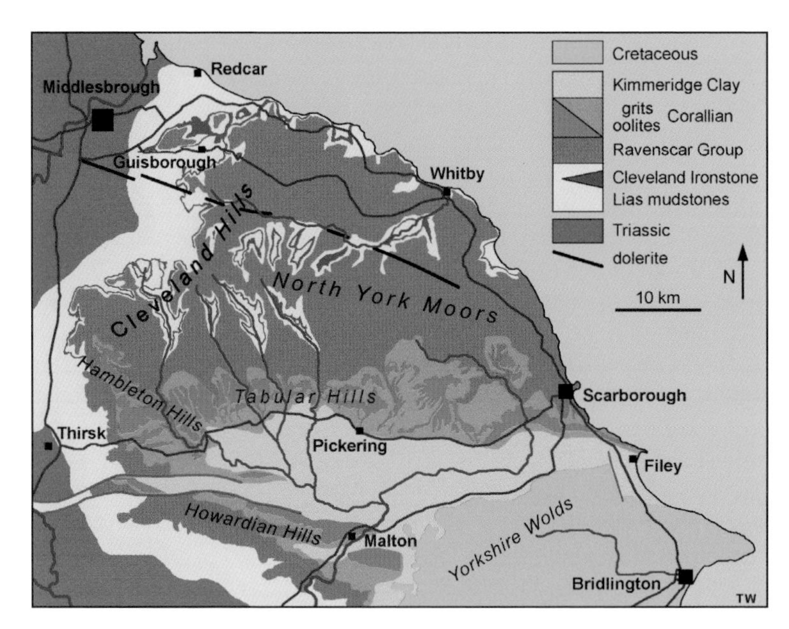

LEFT Main outcrops of the Jurassic rocks across the North York Moors and adjacent hills.

BELOW Stratigraphy of the Middle and Upper Lias in the Cleveland Basin. Thicknesses are generalised and show considerable local variation; the sequence shown here is about 140 metres thick. The Cleveland Ironstone (strictly a Formation) is about 25 metres of mudstone containing relatively thin seams of payable iron ore.

Absence from the Jurassic sequence is largely due to non-deposition in the erosional environment of land above sea level. Levels of both land and sea varied over the millennia, and some phases of localised sediment deposition were followed by their removal by erosion, but such were details. It was rather more significant to Jurassic geology that the Cleveland Basin could subside and fill with marine and deltaic sediments, while the Market Weighton land, whether island or peninsula, kept it largely separate from the shelf seas that extended southwards across most of England.

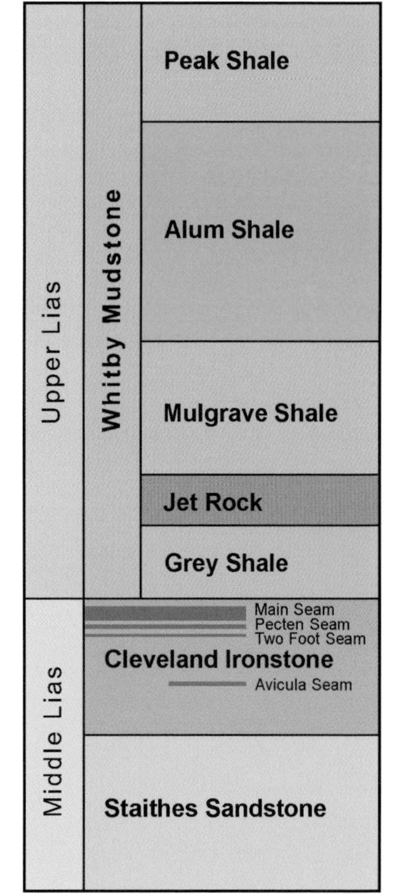

Redcar Mudstone exposed at Boggle Hole, with two beds of fine-grained sandstone conspicuous in the cliff.

Lower Jurassic Beds

As a rock type, lias is an old English term for a hard limestone, typically found as thin, impure beds interlayered with grey clays. Thin limestones and clays form most of the Lower Jurassic sequence in England, and even though clay generally dominates, the whole group is now known as the Lias, a term that has evolved from rock type to stratigraphic unit. Mud is fine-grained silicate sediment composed largely of clay minerals. Buried and lithified in a rock sequence, water is squeezed out so that it becomes clay. Further lithified and hardened, it is known as shale where laminated, or as mudstone if more massive and poorly bedded. Most of the material does end up as mudstone, which is why the stratigraphical names for these units of mudrock include the word Mudstone. However, at and close to outcrop, hard mudstone weathers and softens to revert to more deformable clay, or breaks and splits to form fissile shale. So it is clay or shale that is generally seen at outcrop, which is why many of the beds were formerly known by those names.

Besides these clays, shales and mudstones, the Yorkshire Lias is notable for containing three materials that have supported significant local industries – ironstone, jet and alum shale; the story of local mining forms Chapter 9.

First of the Jurassic rocks to form were the thick clays of the Lower Lias, now formally known as the Redcar Mudstone. The wide desert basins of the Triassic became shallow seas in the Jurassic, with fine-grained sediment supplied from landmasses that are now Scotland, Wales and the Pennines. Hence these shales and mudstones total more than 200 metres in thickness, all formed in about ten million years while the floor of the Cleveland Basin slowly subsided. Well known for their abundance of fossils, they are largely hidden beneath soil along the eastern side of the Vale of York, but are exposed in the cliffs and foreshore at Robin Hood's Bay. The Lower Lias has the only Jurassic outcrop that is continuous into Lincolnshire, as it predates the emergence of the land barrier at Market Weighton.

Three thin beds of ironstone form ribs within the cliff of dark mudstone that dominates the Cleveland Ironstone Formation east of Staithes harbour.

Staithes Sandstone forming the cliffs on both sides of Staithes village (which is largely hidden in this view from Boulby Bank).

Middle Lias Ironstone

Shallow marine sediments continue into the Middle Lias, but are distinguished by being sandier due to rivers on the northern lands carrying increasing amounts of sand into the Cleveland Basin. Some 30 metres thick in the Eston Hills near Middlesbrough, the Staithes Sandstone thins to almost nothing beneath the Vale of Pickering. Nowhere is it a conspicuous feature of the landscapes, and most of its outcrop is covered by glacial till, except at its type locality at the pretty village of Staithes. There, its thinly bedded sandstones are nearly horizontal where they form that splendid rock prow rising behind the old cottages on the left bank of the narrow inner harbour.

Thin beds of ironstone are weathered to a red crust where the Cleveland Ironstone Formation is well exposed during low tide on the foreshore around the headland of Old Nab, east of Staithes.

The next major stage in Jurassic evolution was the greatly increasing occurrence of iron minerals within the sediments of what became the Cleveland Ironstone. This unit of rocks is up to 30 metres thick in the northern Cleveland Hills and in Eskdale, but then thins southwards to almost nothing. It consists mainly of mudstone, but has more than a dozen beds of ironstone scattered through its sequence, and these are dominant in the top six metres. Most of the ironstone is composed of roughly equal parts of chamosite (a green iron silicate), siderite (pale brown iron carbonate) and a mixture of detrital minerals. The iron content reaches to around 30%, so beds more than about a metre thick could be mined – and supported a lucrative industry in the Cleveland Hills for about a hundred years, until the last mine ceased operations in 1964.

Most of the ironstone is oolitic, with ooliths of chamosite set in a matrix of siderite and non-ferruginous minerals. Ooliths (also known as ooids) are sand-grain-sized spheres with internal concentric layers that were formed as wave action rolled the grains around on the shallow sea bed. The iron appears to be polygenetic – like so many of features of geology, it has multiple origins. Much of it was precipitated out of solution in the seawater, generally aided by bacterial activity that influenced the water chemistry. But iron is a very mobile element in an aqueous environment, and further deposition and mineral changes took place during diagenesis – when the sediment was slowly turned into rock after burial deep beneath later sediments. Furthermore, the iron within the seawater probably had mixed origins. Much of it was derived from erosion of the nearby landmasses. Then it is likely that iron was also added to the seawater by hydrothermal vents on the floor of the Atlantic Ocean that was starting to open not far to the west when the American continent drifted away from Eurasia.

Nothing lasts for ever, and environmental changes near and far caused the iron deposition to cease, though only temporarily. Meanwhile, sediment sequences in the shallow seas of the Cleveland Basin reverted largely to the ubiquitous mud and sand.

Upper Lias Jet and Alum

Whitby Mudstone is the formal name for a sequence more than 100 metres thick, most of which appears as shale at outcrop, though it also includes some significant variants. They are poorly exposed inland, but are important features of the coastal strip around Whitby, and have fine exposures in the sea cliffs. The basal Grey Shale becomes gradually darker as it passes up into the Jet Rock, most of which is again finely laminated shale, but contains scattered lumps of jet and is black due to its high content of organic carbon.

Jet is the amorphous, shiny, black stone that can be carved into ornaments and jewellery in a small but significant local industry. It is fossilised wood, so essentially a variety of lignite coal, forming what were originally tree trunks, mostly of araucarians, distant relatives of the spikey 'monkey-puzzle' trees of modern South America. Hence the main occurrence of jet is as isolated lumps scattered through the top few metres of the Jet Rock shales. Rivers on the low-lying, sub-tropical hinterland carried logs to the coast beyond, which then sank into muddy sediments that were accumulating beneath the shallow waters of the Cleveland Basin. Buried by further mud deposition, the water-logged wood was prevented from oxidising, so that it lost its volatiles and was then compressed and distorted into lumps that are structureless, tough, and more than 70% carbon. Today jet is well known as fragments on the beaches around Whitby, where it is derived from erosion of coastal outcrops of the jet-bearing mudstones.

In formal stratigraphical nomenclature, the Jet Rock is the lower part of the Mulgrave Shale, which continues upwards with a high content of hydrocarbons (hence its earlier name of Bituminous Shales), though almost completely lacking in jet.

That is then followed by the Alum Shale, which was quarried extensively in the past (*see* Chapter 9) to produce alum for the textile industry. Alum is the double sulphate of aluminium and potassium, and, as the mineral alunite, is rare in nature except at some volcanic sites. However, the Alum Shale contains the requisite aluminium (in the clay minerals, which are alumino-silicates) and sulphur (in pyrite, iron sulphide, which oxidises to produce the sulphate) for making alum; the potassium is added separately. Alum Shale is therefore little different from many other shales that contain scattered crystals of pyrite, except that it also contains organic carbon (which burns to help the calcining), is low in calcite (which would steal any new sulphate to form gypsum), and has about 3% very fine-grained pyrite (which is the right amount to generate the sulphate); these three factors are the keys that made only the Alum Shale suitable as the raw material to feed the prolonged and complex processing

An old quarry west of Ravenscar once yielded aluminous shale for the local alum works. Part of the Whitby Mudstone, the aluminous shales are now largely hidden behind the bank of shale scree, while the top of the face exposes the caprock of Dogger sandstone that prevented further quarrying into the hillside.

that eventually yielded the precious alum. That industrial process was comparable to hugely accelerated weathering, but natural alunite is absent in the Alum Shale even where it has weathered in the natural environment.

Around the northern and coastal margins of the North York Moors, the Alum Shale reaches to nearly 40 metres thick. However, only 15 metres of this, near the middle of the sequence, was suitable for alum extraction; the lower beds have less organic carbon and more iron carbonate, and the quarrymen avoided them because they are harder; then the upper beds contain too much calcite – indeed some were quarried in the past to make cement.

The top beds of the Whitby Mudstone are named Peak Shale, seen only in a narrow slice east of the Peak Fault (*see* Chapter 4). Across most of the North York Moors it was removed by erosion during a brief interlude of uplift, before the Cleveland Basin subsided yet again and Middle Jurassic sedimentation restarted above the unconformity.

A pyritised Microderoceras***, less than 30mm across, from the Redcar Mudstone at Robin Hood's Bay.***

Fossils of the Lias

Among the most famous fossiliferous beds in Britain, the Lower Jurassic clays, shales and mudstones yield the richest pickings at their two coastal outcrops, in Dorset and Yorkshire. Robin Hood's Bay forms the heart of Yorkshire's fossil coast, though the nearby bays at Staithes and Port Mulgrave, along with Saltwick Bay on the other side of Whitby, are generally the better sites for collecting. At each site, the foreshore exposures are excellent though only accessible at low tide, and the best collecting is always in winter after storm waves have broken fresh material from the cliffs. The days are long gone when weather-beaten specimens of the bivalve *Gryphaea*, or even a decent ammonite, perhaps not complete, could be picked up by a walker on the Yorkshire beaches. Diligence, patience, a hammer and a touch of luck are now needed

ABOVE ***A fine specimen of the ammonite*** Pleuroceras ***revealed by breaking open a nodule in the Whitby Mudstone.***

RIGHT ***Fossil hunters on bedded mudstone of the Alum Shale exposed on the foreshore beneath the East Cliff at Whitby, with Ravenscar sandstone forming the top of the cliff.***

Dactylioceras commune, ***about 70mm across, the most common ammonite in the Whitby Mudstone.***

Hildoceras bifrons, ***with whorls having smooth inner and ribbed outer sections; from the Alum Shale.***

for a casual visitor to collect a good fossil. The professional fossil collectors choose their time, usually in winter, know where to look after any rockfall from the cliffs, and can recognise a block with potential; then many hours of curating can produce some beautiful specimens, especially of clusters of ammonites.

Ammonites are always a delight to find, and the simple ribbed coils of *Dactylioceras* are perhaps the most abundant in the Yorkshire Lias. Also numerous in the mudstones around Whitby, *Hildoceras* is a distinctive ammonite of special local significance as it was named after St Hilda, Abbess of Whitby, during the seventh century; legend has it that she turned a plague of snakes to stone, presumably after coiling them up. Known as snakestones, the fossils therefore feature on Whitby's coat of arms. The stratigraphy of the Lias is identified by its ammonites, of which many species are true markers for individual zones with the succession. Most of the Liassic ammonites are less than ten centimetres across, so specimens can be complete when they are broken out of the nodules that are often collectable. Nodules are rounded lumps of the mudstone that are concretions formed where stronger cementing mineral was deposited between the grains in concentric zones around some nucleus. Commonly, it was a fossil acting as the nucleus that provided the chemical contrast to cause the mineral precipitation. Mostly fist-sized within the Lias, these nodules weather out of the cliffs, and some can be broken up to reveal decent fossils.

An assemblage of the ammonite Paltechioceras ***in a block of Redcar Mudstone about 250mm across.***

Brachiopods are common throughout the Jurassic sequence. They were bivalves that lived on the sea bed, so grew thick, unequal calcite shells, either ribbed or smooth, as defence against predators, and consequently many survived burial and fossilisation. There is also a class of bivalves, within the mollusc phylum; previously known as pelecypods or lamellibranchs, these include the oysters. Most of them have thinner shells that were easily crushed during burial. A notable exception was *Gryphaea*, which had one large shell curved and almost wrapped round a small

Gryphaea incurva, ***the asymmetrical bivalve fossil that is common in much of the Jurassic succession.***

Four vertebrae (each 30mm thick) and six smaller bones from the paddles of an ichthyosaur found in the Whitby Mudstone near Samdsend.

second shell. Usually a few centimetres long, conspicuous growth lines on the curved shell led to its nickname as a 'devil's toenail'. Along the Whitby coast these fossils are commonly weathered out of the mudstones and rolled around by wave action, when they become smoother and more rounded; but they are strong enough to survive, so can still be collected on some of the beaches.

The coastline of Liassic rocks alongside the Yorkshire Moors is sometimes known as the Dinosaur Coast, though this is something of a misnomer. Dinosaurs lived on land, so when they died their remains were generally scavenged, broken up, eroded away or simply weathered to leave little trace. Their fossils survive mainly where the animals were buried in deltaic sands, so they are lacking in the marine rocks of the Lias. Within the Middle Jurassic Ravenscar sandstones (*see* below), delta-top beds near the top and bottom of the sequence contain dinosaur footprints, though bones are rare. Many sets of footprints have been exposed on the foreshores north and south of Scarborough, but situations change with on-going marine erosion of the sites.

However, the marine Lias is famous for its large vertebrate fossils, notably in the form of plesiosaurs and ichthyosaurs. These giant reptiles were the predators of the sea, up to three metres long, streamlined with powerful paddles and long, toothed jaws, and their fossils are particularly abundant in the Alum Shale. Individual teeth, vertebrae and disarticulated bones can be found in the coastal exposures, though discoveries of complete skeletons have largely been the preserve of quarrymen. During the 1800s, coastal shale quarries that fed the alum works at Ravenscar and Loftus yielded magnificent specimens that are now housed in museums across the country. Large specimens turned up in such numbers that geologists visiting from afar during the early 1900s commented on fossil vertebrae being used to line garden paths at homes around Whitby.

Skeleton of an ichthyosaur from the Liassic mudstones (this is a plaster cast of the fossil, in a block 700mm long).

Middle Jurassic Sandstones

Nearly ten million years of deltaic deposition created the sandstones that are the bedrock, the essence, the defining feature, of the North York Moors. They dominate the sequence known as the Ravenscar Group, where they are strong enough to form the high ground of the Moors and the Cleveland Hills characterised by classic heather moorland on their poor quality, sandy soils. The same sandstones form parts of the Howardian Hills, but somewhat lower in altitude these are largely developed into productive farmland.

Around 200 metres thick in total, these sandstones have previously been called the Estuarine or Deltaic Series, because they were deposited within the coastal and near-offshore zones of a broad delta that was fed by rivers pouring in from the northwest. Open sea lay eastwards, north of the contemporary Market Weighton land. The Ravenscar Group is dominated by these sandstones, and is sub-divided into a sequence of Formations (as in the table below). The three thick deltaic formations include large amounts of siltstone and mudstone, along with thin coal seams, plant beds and bedding planes with dinosaur footprints, all reflecting intervals when delta-top environments prevailed. Coal seams are only thin, but those near the top of the Cloughton Formation were worked long ago in hundreds of small pits across the Moors. Less than 50 metres above the coals, the Moor Grit Member is a particularly strong bed of sandstone, about ten metres thick, that gained its name from defining some areas of the finest high moorland, even though its outcrops are actually not that extensive.

Within the Ravenscar sandstones there are four relatively thin formations that record marine incursions when slow regional subsidence briefly outstripped the rate of sediment accumulation. Lowest of the four is the Dogger, with up to ten metres of coarse sandstone that is commonly calcareous and locally pebbly. It includes beds of ironstone with chamosite and siderite, similar to the Cleveland Ironstone seams. Furthermore, its outcrop around the rim of Rosedale was famed for the remarkable magnetic ironstone

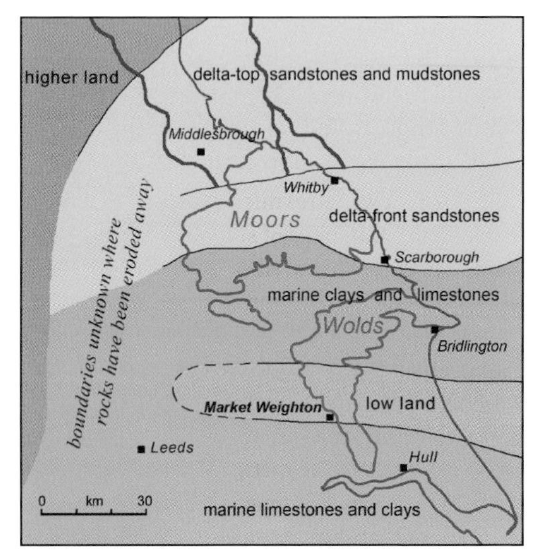

ABOVE ***A hugely simplified palaeo-geography of the Jurassic sea within the Cleveland Basin. Features varied considerably through the 50 million years of Jurassic time, but were dominated by sediment from land in the northwest and open sea on both sides of the small Market Weighton landmass.***

LEFT ***Bedded sandstones of the Ravenscar Group exposed on the neck of land reaching out to Scarborough Castle.***

that was mined in the late 1800s. This is of limited extent as it occurs only within some channel deposits lying within the Dogger, but its iron content of up to 50% is a significant improvement on the 30% or less that is typical for ores of the Lias.

There is still some debate over the formation of this valuable iron ore, and it remains unresolved because the orebodies cannot be reached for study in the abandoned and collapsed mines. Current thinking has an extensive sea-bed deposit of the Dogger ironstone that was reworked beneath shallow waters to fill submarine channels that had been scoured through it and into the underlying muds. The problem is that magnetite (the double oxide of iron, Fe_3O_4) is not normally found in the sedimentary environment, so it is likely that chemical changes of the iron minerals were induced at a later stage, by groundwater flowing through the sediments while they were being lithified and turned into solid rock.

The Eller Beck Formation consists largely of shales just a few metres thick, with marine fossils, but includes thin beds of siderite ironstone with limited lateral extents. Above this, the Whitwell Oolite is a pale, shelly limestone that only occurs in the southern part of the Cleveland Basin, notably in the Howardian Hills where it is nearly ten metres thick. Formed during the fourth of the marine incursions, the Scarborough beds reach to around 20 metres thick with coarse sandstones, sandy limestones and sideritic mudstones that show lateral variation round the edges of the Moors.

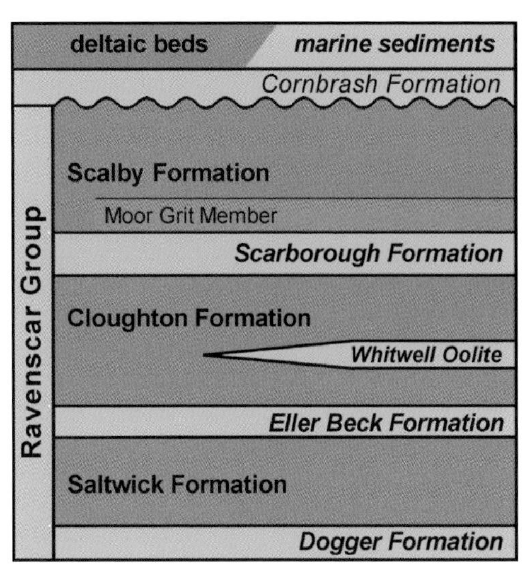

Major stratigraphic units within the Ravenscar Group, dominated by the stronger sandstones that form the North York Moors.

Quarries west of Aislaby work into their thick bed of sandstone overlain by thin-bedded sandstone and darker mudstones, all within the Saltwick Formation.

Dogger and Saltwick sandstones on the Kettleness headland, where the Alum Shale was extensively quarried from the slopes below.

A slab of Moor Grit, from its outcrop on Spaunton Moor, now upended as the Millennium Stone on Danby Moor.

A further marine incursion is represented by the Cornbrash, a few metres of sandy limestones that lie directly above the Ravenscar Group. An unconformity lies between the two beds, indicating a brief interlude of uplift and localised erosion of the uppermost Ravenscar beds, before the Cornbrash sea swept in. There is no angular contrast, so this is technically a disconformity. And although conformable with overlying Upper Jurassic beds, the Cornbrash is placed in the Middle Jurassic on the basis of stratigraphic relationships elsewhere.

A fossil brittle star in a block of sandstone recovered from the cliffs west of Skinningrove.

Upper Jurassic Beds

Two great units of clay, the Oxford and the Kimmeridge, were deposited across huge areas of the Jurassic seas, in what is now the south of England. Both also extended northwards round the Market Weighton landmass, to where they now underlie lowlands in East Yorkshire. Between them lie the Corallian beds that form further significant parts of the Yorkshire landscape.

Clays are poorly exposed in any temperate landscape, as they weather to deep soils and underlie the lowlands where sediments accumulate over them. The Oxford Clay is no exception. Generally around 30 metres thick, it may better be described as a mudstone, and is poorly exposed except at the coast around Cayton Bay. Underlying the clay, the Osgodby sandstones include strong beds that were formerly known as Kellaways Rock, and also the strong Hackness Rock. The Oxford Clay is important within the landscape because it forms a weakness between the strong Ravenscar sandstones beneath and the equally strong Corallian grits above. Consequently, its sinuous outcrop lies along the foot of the dissected escarpment that forms the northern edge of the Tabular Hills where they overlook the Moors rising gently to the north.

In contrast, the Kimmeridge Clay, consisting of more than 200 metres of dark clays and shales, underlies almost the entire Vale of Pickering. These weak clays are the reason behind the existence of this great sweep of lowland, and also the great indent of Filey Bay where the same bed meets the coastline. But nowhere is this bedrock exposed: it is covered by lake sediments throughout the Vale and by glacial till in the coastal zone, and its presence is only known from boreholes.

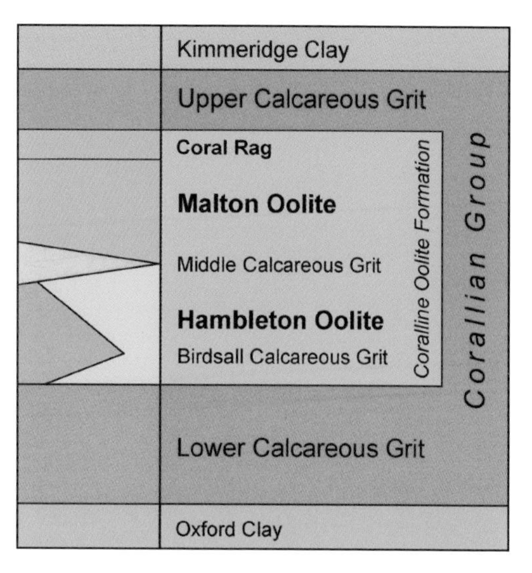

Stratigraphy of the Corallian beds, with the oolitic limestones and calcareous grits all varying in extent across outcrops in the hills both north and south of the Vale of Pickering. The oolitic limestones generally form about half of the total Corallian thickness of around 100 metres.

Corallian Beds

Between the two clay beds, the Corallian Group consists of around 100 metres of limestones and calcareous sandstones. The former are the Hambleton and Malton Oolites, and the latter are the Calcareous Grits. Together, these resistant rocks form the low, dissected plateaus of the Hambleton Hills and the Tabular Hills, and also extend eastwards to form the

Newbiggin Cliff stands 60 metres tall, west of Filey Brigg; grey Oxford Clay is overlain by brown Lower Calcareous Grit beneath a cap of glacial till that is shrouded in greenery.

Lower and Birdsall Calcareous Grits exposed in the cliffs along the north side of Filey Brigg from beneath a thick cover of glacial till.

high ground of Dalby Forest, where the grits predominate. The Lower Calcareous Grit survives to form large parts of the headlands of Filey Brigg and Scarborough Castle, and also forms much of the northern half of the Howardian Hills alongside the Malton Oolite. Across most of the Tabular Hills, the full sequence is in place with the two oolites and the three grits, though the lower part is more variable further south. In the Hambleton Hills the eponymous oolite is present, whereas the Malton Oolite is missing. Further east, along the Howardian Hills, the Malton Oolite is thick and extensive around its namesake market town. Also, the Birdsall Grit becomes thicker, eventually replacing the Hambleton Oolite, and converging with the Lower Calcareous Grit to form a thick grit unit that is directly overlain by the Malton Oolite. Such are the complications of stratigraphy where shifting coastal environments created lateral changes between rock units of the same age.

The Corallian beds were all formed in a shallow tropical sea where a wealth of marine life provided shell debris to form the limestones. Incursions of quartz sand were produced by erosion of land away in the north and west, to

Gently dipping beds of Birdsall Grit are exposed only at low tide on the headland of Filey Brigg.

form each of the beds of Calcareous Grit. These only really warrant being known as grit because they are topographically resistant, as they are mainly fine-grained sandstones with high proportions of carbonate. It appears that many of the silica grains within these grits originated as the skeletal spicules within sponges. The gently dipping slabs of Birdsall Grit exposed at low tide on Filey Brigg are remarkable for their abundance of *Thalassinoides* burrows. These were made by relatives of shrimps and crayfish that fed on nutrients within the shallow-water sand, sometimes making many burrows radiating out from a small central chamber. Abandoned burrows became filled with sand that appears to be more resistant than the host sand, so the burrows weather out in spectacular form on the foreshore exposures.

An oolitic limestone consists of ooliths (also known as ooids), sand-sized, spherical concretions of calcite that formed around fragments of shell debris rolled around on the floor of a shallow sea. The Corallian oolites appear as fine-grained fossiliferous limestones, as they contain relatively few small ooliths in a matrix dominated by shell debris. Scattered patches of coral reef were also a feature of the Corallian sea, and these eventually extended to form the Coral Rag at the top of the Malton Oolite. Locally reaching ten metres thick, the Rag is a distinctive limestone composed of coral debris and broken shells of bivalves, gastropods and echinoids, with an upper layer that is rich in corals in growth position.

Multiple sand-filled burrows of Thalassinoides ***exposed in a block of Birdsall Grit on Filey Brigg.***

Limestone deposition in the clear warm sea of the Cleveland Basin ended with a final incursion of Calcareous Grit. Subsequently the waters turned muddy, when the great mass of Kimmeridge Clay was derived from weathering of land areas where the hills had been worn down and sluggish rivers prevailed. This was followed by a phase of regional uplift that left the whole of Yorkshire as dry land, which persisted for about 40 million years. Consequently the uppermost Jurassic succession, and most of the Lower Cretaceous, is entirely missing, though the scene did change later in the Cretaceous.

Lateral variations within the Malton Oolite and Coral Rag exposed in a wall of the abandoned Wath Quarry, in the northern flank of the Howardian Hills.

CHAPTER 3

Cretaceous Rocks of the Wolds

Within Britain, the Cretaceous rock sequence is dominated by the Chalk, and nowhere more so than beneath the Yorkshire Wolds, where nearly all the Lower Cretaceous beds are absent. This is because they were never deposited on the Market Weighton Block, where dry land persisted from Jurassic times into the early Cretaceous.

North of the Market Weighton land, some marine sediment was deposited during the Lower Cretaceous. This is dominated by the Speeton Clay that survives in a strip beneath the southern part of the Vale of Pickering, until it is cut out west and south of Malton by the Chalk unconformity. It has almost no inland exposure, but can be seen at low tide in the low cliffs east of Speeton at the southern end of Filey Bay.

Lying directly above the Speeton Clay, the Red Chalk is an enigmatic bed, well known in Norfolk, but difficult to see in Yorkshire. A sliver within the cliffs over Speeton beach is often obscured by fallen debris, and there is a conserved exposure in the old Rifle Butts Quarry, in Goodmanham Dale, two kilometres east of Market Weighton. It is actually the basal few metres of the Chalk, but is distinguished by its brick-red colour due to a tiny content of red iron oxide in the form of hematite. Other than that hematite content, much of it is a typical chalk, the same as the huge thickness of white rock above it, though the lower part locally has more pebbles and impurities. It appears to have been deposited in a shallow sea or even in the intertidal zone. The iron mineral could have been derived from a desert environment

The chalk cliffs of Flamborough Head, with their cap of glacial till hidden in the greenery.

on the Market Weighton landmass, or is more likely to be related to volcanic activity where the nearby Atlantic Ocean was opening up on a divergent plate boundary.

After a long interval without deposition of sediment across much of the Wolds region, sea levels rose during a phase of global warming later in the Cretaceous. Only then did the low-lying land on the Market Weighton Block disappear beneath the waves, and the sediments of the Chalk were deposited in shallow seas across the whole of eastern Yorkshire.

Chalk: the great white rock

The word 'chalk' has two meanings. It describes a particular type of soft limestone, so is written as chalk without an initial capital. It is also a major stratigraphic unit in the Cretaceous, so is then written as Chalk with an initial capital, and is commonly prefixed by a group name, such as Flamborough Chalk, which is the main unit in the Yorkshire Wolds.

The well conserved exposure in the old Rifle Butts Quarry. White chalk overlies less than a metre of Red Chalk, which sits directly on Redcar Mudstone that forms the floor in front of the face. The old quarry face is kept clear of invasive plants, is protected by the cover of wire mesh, and is under a rain shelter.

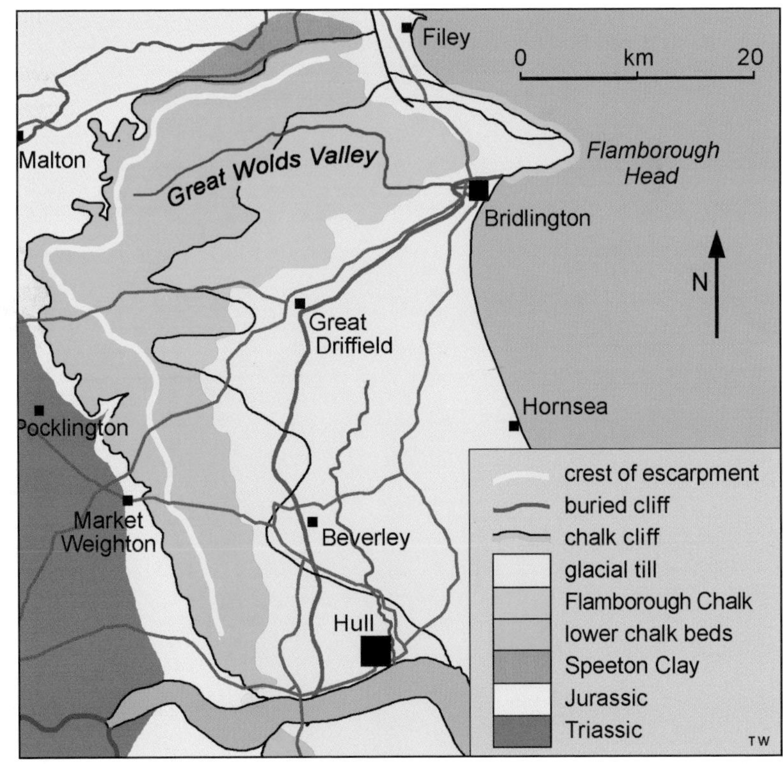

Outcrop of the chalk, which marks the extent of the Yorkshire Wolds. East of the buried cliff, the top of the chalk is below sea level. The line marking the crest of the Wolds escarpment is smoothed, with the steep scarp face to its north and west.

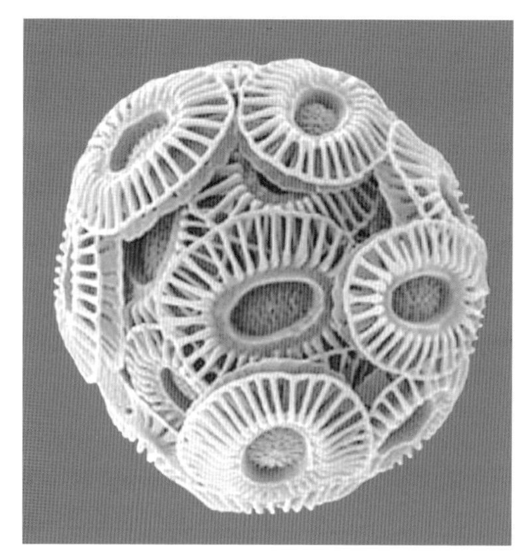

A single coccosphere, less than one hundredth of a millimetre across, consisting of many of the coccolith plates, themselves skeletal in structure, that are the major component of chalk.

Chalk is a remarkable rock, almost entirely made of calcium carbonate in the form of calcite. It is a very pure limestone, but is poorly lithified, so it can be soft and friable, though it varies considerably in strength. The softest chalk can be used to write on a blackboard (though commercially available sticks of chalk are made of gypsum); most is rather harder, though nowhere is it as strong as the Pennine limestones. It is made almost entirely of coccoliths, which are calcite plates, each just a few microns (a few thousandths of a millimetre) across in the shape of skeletal discs. These were the components of almost spherical skeletons (known as coccospheres) formed inside single-celled algal protozoans that are known as coccolithophores, and these could grow to be as large as a fiftieth of a millimetre in diameter. Unimaginable numbers of these protozoans floated in the warm seas on the continental shelf until their skeletal plates rained down to the seabed, to form a rather special variety of calcareous ooze. Buried by more of the same, this recrystallised to produce chalk rock.

Most of the Yorkshire chalk is a relatively hard and thinly bedded variety, rather different from the softer and more massive chalk of southern England. This is a result of pressure-solution of carbonate shells, followed by deposition as calcite cement in the pore spaces of the sediment. Essentially just a greater

Gently rolling farmland that is typical of the Wolds, with the underlying chalk bedrock exposed in a large abandoned quarry once worked for lime in the northern flank of Burdale.

Micraster, ***one of the most common echinoid fossils in the Chalk; this specimen is 50mm in diameter.***

degree of lithification, this is normally due to greater depths of burial, but obscure chemical processes account for the local variations within the Wolds chalk.

During Late Cretaceous times, when the English Chalk was formed, worldwide temperatures were at a major high. There were no polar ice-sheets, and sea level was about 200 metres higher than it is today. The sea spread across most of England, while the Pennines and Lake District joined to Wales and Scotland, forming an island of old hard rocks. The whole world was warmer, and Britain was further south before continental drift brought it to its present latitude. So its climate was sub-tropical, though most of it was under water. Coccolithophores thrived in those warm seas for about 30 million years, so that their calcite skeletons accumulated to thicknesses of around 500 metres. This chalk sediment was notably pure. Thin beds of chalk marl, many only a few centimetres thick, contain clay that could have been derived from distant lands, though at least some of it appears to be ash from volcanoes that were active while the Atlantic Ocean was widening, and were not as far away as those on the Mid-Atlantic Ridge today. Very fine volcanic ash may also have contributed to microscopically thin layers of dust that form some of the bedding planes within the chalk.

Minimal sediment from land, and occasional volcanic eruptions, were all that interrupted the rain of shell debris that sank to the sea floor as the remains of billions of free-swimming animals both small and large. The smaller fauna created the chalk itself, whereas some of the larger became more visible fossils. Echinoids (sea urchins) and crinoids can be found, but can be difficult to extract from the rock. Ammonites and bivalve molluscs also occur, but the easiest to find are the belemnites; their solid calcite sections at the end of the shells' living quarters survive to form the strong, bullet-shaped fossils that are commonly weathered out of the weaker host rock.

Though often difficult to find, the larger fossils are the key to the stratigraphy of the Chalk. The Wolds Chalk is divided into four formations (*see* the table below). Thickest is the Flamborough Chalk, which occupies by far the most extensive outcrop and also

geological units		local beds	Ma
Upper Cretaceous	Senonian	**Flamborough Chalk**	70
		Burnham Chalk	
	Turonian	**Welton Chalk**	
	Cenomanian	**Ferriby Chalk**	
Lower Cretaceous	Albian	**Red Chalk**	100
	Aptian and lower	**Speeton Clay**	145

Stratigraphy of the Cretaceous rocks exposed in the Yorkshire Wolds. Relative thicknesses are indicative but not precise; the Red Chalk is generally less than ten metres thick, whereas the Flamborough Chalk is more than 200 metres thick. The top of the chalk in Yorkshire dates to about 70 million years ago, which was nearly four million years before the end of the Cretaceous.

forms most of the cliffs round the eponymous headland. In contrast, the Bempton Cliffs are formed of the Welton and Burnham Chalks, which extend into narrow outcrops along the scarp slopes forming the northern and western fringes of the Wolds. At first glance, the entire Chalk appears to be almost tediously uniform, but there is a notable variable: the Welton and Burnham Chalks both contain numerous flints, whereas the Flamborough Chalk has none, as does the thin Ferriby Chalk at the base of the succession.

The great warmth of the Cretaceous ended with a massive phase of worldwide cooling. Led by long-term astronomical cycles, this progressed slowly through more than ten million years. Midway it was interrupted, or superimposed, by the effects of the gigantic

Flint

Flint is microcrystalline silica. Along with chert and agate, flint has the chemistry of quartz but lacks the ordered atomic structure that can form large crystals. It is hard and chemically resistant, so it survives weathering to form many of the pebbles and cobbles on the beaches south of Flamborough, and is also common in the soils and glacial tills of the Wolds and Holderness. Black, white or grey, flint occurs only within chalk, mainly as scattered lumps known as nodules; these commonly have a thin white crust, known as a cortex, of porous silica in which tiny air bubbles reflect the light.

The silica that forms flint was derived from marine organisms that had siliceous skeletons, including diatoms, radiolaria and many of the sponges. These lived as minority groups in the Chalk seas, and their skeletal debris became a minor siliceous constituent disseminated through the seabed ooze that was formed largely of carbonate. Then some time during the long, slow process of burial and lithification, the silica went into solution, migrated, and nucleated to form the nodules, leaving behind a chalk that is now almost pure carbonate. There is some debate over exactly how and when this occurred. Bacterial activity produced acids needed to mobilise the silica, a process that could have started in the soft sediment. But then some of the precipitation of the silica can be seen to have occurred in joints that developed later in lithified rock.

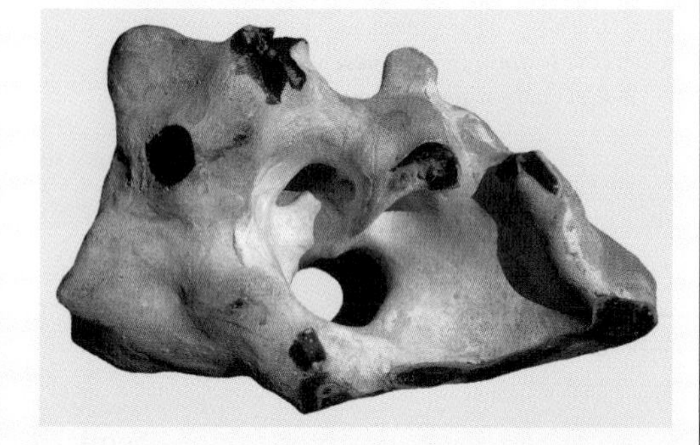

Convoluted shape in a nodule of flint 20cm long. The white crust is thin, and the solid flint inside is nearly black, as seen where projections have been broken off.

Nodules of flint along a bedding horizon in chalk exposed in Thornwick Bay.

Many flint nodules formed around fossils or shell fragments that were the focus for precipitation. Some of the larger nodules, extending up and down across the bedding, appear to have replaced the fills in animal burrows. And some flint achieves almost tabular form where it nucleated around clay impurities that defined bedding planes. Flint appears to be polygenetic, with alternative means of formation that are still not fully understood, but it remains a distinctive feature of chalk terrains such as the Yorkshire Wolds.

Eroded, weathered and pitted belemnites from the Chalk. Each of these is a rostrum, or guard, that is the solid calcite at the rear end of the hollow cone in which the animal lived. These therefore survive well as fossils; each is about 70mm long.

volcanic eruptions creating the basaltic Deccan Traps of peninsular India, and also by the giant meteorite impact at Chicxulub on the Caribbean coast of Mexico. The meteorite impact marks the end of the Cretaceous, whereas the Deccan volcanoes straddle the boundary because they were active for about a million years. All, or some, of these caused a mass extinction of marine life, and also the final demise of the dinosaurs around 66 million years ago.

The cooling also saw a huge decline of sea level, so that the whole of Yorkshire became land, which reached 100 km out beyond the present coastline and south as far as Norfolk. Never again was Yorkshire submerged, so there were no more sediments except those left by rivers and glaciers, and the county has no sedimentary rocks above the Chalk.

Cretaceous rocks were later crumpled to form the great mountain ranges of the Alps, but this far north, those same earth movements created only modest undulations of the rock sequence. Now the Chalk forms its long outcrop from Flamborough Head to the Dorset coast. West of that it has been eroded away, but eastwards it dips gently beneath Holderness, beneath most of eastern England, and beneath much of northern Europe.

Nearly horizontal, bedded chalk exposed in the high cliffs at Bempton, north of Flamborough Head.

CHAPTER 4

Rock Jigsaw

Within the Jurassic and Cretaceous rock sequences of eastern England, geological structures are mostly gentle and simple, because the region lies on a chunk of relatively stable continental plate. Since before Jurassic times, it has never been caught up in powerful tectonic movements, when the nearest plate boundaries have been divergent in the middle of the Atlantic Ocean and convergent through the Alps and Mediterranean. So the geology of the Moors and Wolds can almost be simplified as a single sequence of rocks dipping gently to the south or east.

Both the Moors and the Wolds have the North Sea to their east and the Triassic lowlands to their west. A broad band of soft mudstones and friable sandstones underlies the Vale of York, together with continuations southwards beneath the Trent Valley and northwards beneath the lower Tees Valley. All the Triassic beds, and the underlying Permian limestones, dip gently eastwards, off the Pennine Anticline and into the North Sea Basin.

This takes them far beneath the North York Moors, where the Permian limestones transgress into sequences of evaporite minerals deposited in a shallow inland sea that was losing its water to solar-powered evaporation. This sea, the Zechstein, extended all the way to Poland. Its sediments now lie at depth except where they rise to outcrop against the Pennine Anticline, but there the basinal evaporites are replaced by coastal limestones. This is unfortunate, because the evaporite minerals are of significant value but can only be extracted by way of very deep mines beneath the Whitby area (*see* Chapter 9).

The regional dip of all rocks older than the Chalk has been gently distorted by the Market Weighton Block. This structurally high area, described by geologists as a positive area, appears to be due to the presence of a granite at depth, though this has not been reached by any borehole. Granite is slightly lighter than most other rocks deep within the crust, so it tends to float upwards. After probable intrusion during

The escarpment of the Yorkshire Wolds, in shadow on the right, with the chalk dipping gently away to the right, and overlooking the eastern end of the Vale of Pickering.

ABOVE ***Beds of the Birdsall Grit dipping a few degrees to the south where exposed on Filey Brigg.***

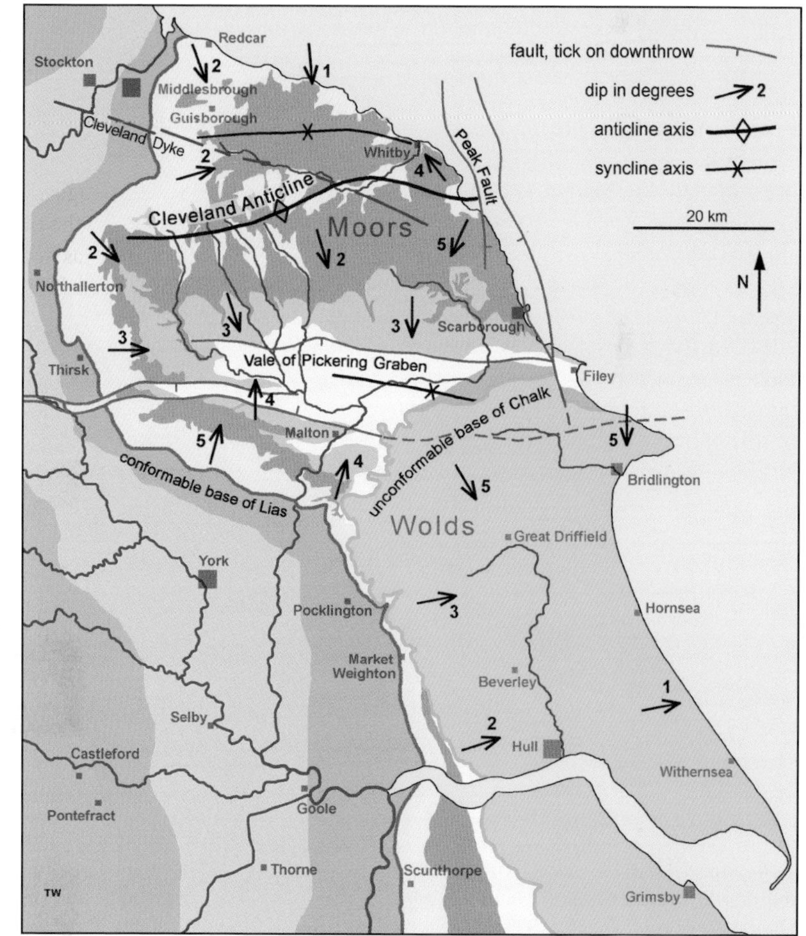

The major structural features within the rocks that form the North York Moors, the Yorkshire Wolds and adjacent areas. Colours of the geological units are paler versions of those in the map and stratigraphical column on page 8.

Robin Hood's Bay, carved into the coastline of the North York Moors where the softer Redcar Mudstone rises to outcrop over the gentle crest of the Cleveland Anticline.

Triassic times, this kept the Block as a landmass throughout the Jurassic, so that it lacks the sequence of sediments that accumulated so conspicuously in the Cleveland Basin to its north. In mid-Cretaceous times, the crust was locked into immobility, perhaps due to gentle tectonic compression, and the Chalk sea could extend unbroken across and beyond the Block, which now lies buried beneath the Wolds.

Folds in the Cleveland Basin

In contrast to the Market Weighton Block, the neighbouring Cleveland Basin was an area of almost continuous slow subsidence throughout Jurassic times. The sequence of shallow marine sedimentary rocks around 700 metres thick indicates the scale of this subsidence, which took place over about 40 million years. Both north and south, the young sedimentary rocks of this basin were effectively trapped between blocks of older, stronger rocks. To the south, the uplifted Market Weighton Block was a well-defined resistant unit, though the north was rather remotely constrained by a Scottish block that extended far out into the North Sea.

So, when much of northern Europe was gently squeezed by north-south tectonic compression during the mid-Cretaceous, the rocks within the Cleveland Basin were buckled into a series of gentle folds. The dominant structure thereby created is the Cleveland Anticline. This is roughly aligned with the highest ground of the North York Moors and the Cleveland Hills, but it is much dissected and the topography is essentially erosional and not structural. The anticline does, however, account for the inliers of Redcar Mudstone along the floors of lower Eskdale and the southern tributaries of the upper dale, along with the broad sweep of Robin Hood's Bay cut into the same mudstones. North of the anticline there is the most gentle of synclines before the main sandstones rise towards their escarpment almost along the coast. South of the anticline, the matching syncline accounts for the sweep

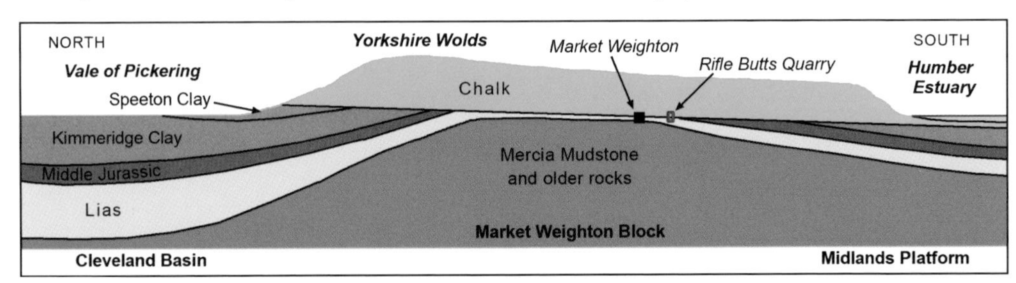

A generalised profile across the Market Weighton Block, drawn close along the western side of the Wolds. The structure is generalised and faults are not shown, notably where a system of small step faults define the northern edge of the underlying Block. The thin layer of alluvium and glacial till in the Vale of Pickering is not shown. The Red Chalk is too thin to appear, lying at the base of the Chalk, and exposed in the Rifle Butts Quarry. The profile is about 50 km long, with vertical scale exaggerated by a factor of twenty.

of the Howardian Hills, though its axial zone is dominated by faulting along the Vale of Pickering. Then the eastern ends of these southern structures are lost beneath the unconformable blanket of the Chalk.

Patterns of tectonic stress change over time. The Cretaceous north-south compression that modestly crumpled the Jurassic rocks was followed by an environment of mild north-south tension that developed during the post-Cretaceous Palaeogene period. This allowed fissures to open as magma invaded from the northwest to form the Cleveland Dyke. Neither of these tectonic situations was a local feature. Both were outlying, marginal features of larger plate movements, with Africa pushing northwards into Europe, and America moving away to the west. In the big picture, eastern Yorkshire's folds and the North Sea Basin were just side effects of the opening of the Atlantic Ocean and the rising of the European Alps.

The Cleveland Dyke

A dyke is a vertical, or nearly vertical, sheet of igneous rock formed where magma was intruded beneath ground level where it flowed into a fissure in the bedrock. Most commonly of dolerite, with the same composition as basalt, dykes are typically a metre or so thick and may extend for a kilometre or so. The Cleveland Dyke is exceptional in that it is mostly about 25 metres wide, and extends for 430 km from a source beneath the volcanic complex on Scotland's Isle of Mull as far as its farthest outcrop on Fylingdales Moor. From there it can be traced westwards for most of the way across the Cleveland Hills to just west of Roseberry Topping, beyond which more scattered outcrops trace across the Pennines and southern Scotland. Parts of the dyke outcrop are broken into *en echelon* segments, but these must be connected at depth.

Abandoned quarries along the outcrop of the Cleveland Dyke form a linear feature across Goathland Moor. Parts of the line are underlain by large underground mine workings that are no longer accessible.

The dyke rock is a quartz-dolerite, with tiny white crystals of feldspar in a dark groundmass, so it is the equivalent of a tholeiite lava, not a quartz-free basalt. Both its composition and its age, at about 58 million years, match with some of the lavas on Mull.

The dyke is a truly remarkable expression of crustal extension, which must have been related to the stretching of the North Sea Basin as a marginal feature of the plate divergence that saw the opening of the Atlantic Ocean. The magma filled the gap in a tension zone, and 25 metres is a significant displacement. There is no known magma source beneath the North York Moors, and it appears that the magma flowed all the way from Mull, filling the fissure as it opened. Considerations of magma viscosity suggest that this all happened within only a few days. Such rapid flow, and rapid cooling, accounts for the minimal baking of wall rocks, reaching only a few metres from the dyke. As a geological intrusion, the Cleveland Dyke is exceptional rather than typical, and as the one igneous rock among many sedimentary rocks it is a complete anomaly within the Moors and Wolds of Yorkshire.

Faults in the Moors and Wolds

The crust of Planet Earth is a very thin layer of rather brittle rock floating on an interior mass that is hot, deformable and constantly on the move – albeit extremely slowly. It is hardly surprising, therefore, that rocks seen at the surface, except for the softest and most plastic clays, are extensively fractured. Joints are nearly ubiquitous in rocks; created by stress and often opened by weathering, they have negligible displacements along them. In contrast, faults are grander fractures where adjacent blocks of rock have been displaced by anything from a few centimetres to many hundreds of metres.

Virtually the only terrains that are not faulted are units of young rocks that sit on top of solid blocks of strong basement, which resist lateral forces imposed by plate tectonics. This applies to the Chalk of the Yorkshire Wolds. It lies on a basement block known as the Midland Platform, with the Market Weighton Block forming its upturned northern edge. Both basement and the Wolds Chalk are now tilted very slightly down to the east, but the great chunk of basement has effectively sheltered its overlying Chalk, so that it has very few faults within it.

Scarborough, with the high ground on the Castle headland formed by a remnant of sandstones and grits separated from the mainland by a fault; this is just one element of the fracture system associated with the Peak Fault.

LEFT Faulted rock in Selwicks Bay, with bedded chalk horizontal on the left, gently dipping chalk on the far right, and steeply dipping chalk in the centre where the right-hand block has been dragged upwards against the fault.

BELOW Calcite crystals in a small vein cavity within the fault zone that is exposed in Selwicks Bay; the yellow pencil is 80mm long.

The same cannot be said for the Jurassic rocks that have suffered deformation within the Cleveland Basin. However, faults are widely scattered across the Moors and are barely recognisable in the topography. One exception is the Whitby fault, which guides the outlet of the River Esk, and also determines the contrast between the East and West Cliffs. East Cliff is a great block of Ravenscar sandstone topped by the remains of Whitby Abbey 50 metres above the shore, whereas the West Cliff has the sandstone down-faulted by about 20 metres and capped by thick glacial till.

East of Whitby, a suite of fractures extends along the coast as far south as Filey. The best known of these is the Peak Fault, at the southern end of Robin Hood's Bay, because east of it there is the only exposure of the Blea Wyke Beds, on and around the headland of the same name. These sandstones lie immediately above the Whitby mudstone, but are not seen west of the Peak Fault. It had long been thought that they were formed of sand caught in a narrow fault-guided trough that acted as a sediment trap. However, current opinion has the Peak Fault as a transcurrent feature that displaced the sandstones from a contrasting sediment zone probably far to the north.

The Scarborough Fault is another within the Peak Fault system. Again it is difficult to recognise the nature of the fault movement, but the effect has been to leave an isolated block of strong Lower Calcareous Grit to form the headland bearing Scarborough Castle east of the fault. North and south of the headland the fault had been crossed by marine erosion, so the North and South Bays are set back into mudstones and weak sandstones that are less resistant components of the Ravenscar Group. The faults continue southwards to where they offset the Chalk and create a rather abrupt end of the high Downs west of Hunmanby.

Between Moors and Wolds the Jurassic rocks are rather more faulted within the Vale of Pickering and the Howardian Hills. The latter form what is essentially a double escarpment, with the Corallian grits along the northern side overlooking the Ravenscar sandstones along the southern. But both are broken by so many minor faults that they have lost much of their impact as landscape features. The faults are all mere components of a major zone of east–west fractures that is rooted in the basement. Some of them define the southern edge of the Vale of Pickering, the sweep of lowland that separates the Moors from the Wolds.

A view south across the Vale of Pickering, with the Wolds along the skyline.

The Vale of Pickering

Besides being a low within the topography, the Vale of Pickering is also a low in the geological structure, which is made more complex by combining both folding and faulting. It can be described broadly as a graben, in that it is a block lowered between parallel faults, created in a zone of crustal tension where a wedge of rock subsided between a pair of steeply inclined fractures. Its southern boundary fault is one of those within the suite known as the Howardian Flamborough Fault Belt. Numerous, mainly small faults are exposed in the Howardian Hills, but continuations to the east are lost beneath the unconformable Chalk. Few of the faults extend up through the Chalk, and most of those remain unseen beneath the soil cover, though some are exposed in the cliffs around Flamborough, notably in the eastern Bempton Cliffs and in Selwicks Bay. The faults continue off-shore, and the coastal exposures have proved useful for geologists trying to understand their role as fluid pathways in the North Sea oilfields.

Although the northern edge of the Vale lies along the foot of the Tabular Hills, the topography is largely defined by faults that truncate the gentle dip-slope. These mark the northern edge of the graben. They lie north of many step faults within the bedrock, which is hidden beneath the Vale's lake sediments, and are known almost entirely from boreholes. The synclinal structure is better developed towards the east, where it almost replaces the graben, before it, too, is lost beneath the Chalk cover.

At the western end of the Vale a separate fault forms the northern margin of a much narrower graben, which reaches through the Jurassic hills to create the Coxwold Gap. The surface expression of a graben is a rift valley, and the section containing Ampleforth College is a fine example, with its steep, wooded slopes rising on each side of a mostly flat valley floor.

View across the Ampleforth rift valley, with the fault-guided northern margin against the Hambleton Hills, which lie shrouded in trees beyond the stone buildings of Ampleforth College and the Benedictine monastery.

CHAPTER 5

Erosion that Shapes the Terrain

Landscapes in eastern Yorkshire have evolved over a few million years of relentless erosion. The land has been there for much longer – about 65 million years since the Cretaceous sea retreated, having seen the deposition of the last of the carbonate sediments that became the Chalk. From then on the land was eroded while it was rising very slowly – partly in isostatic response to the removal of crustal load by the erosion itself. Many hundreds of metres of rock thickness were eroded from the North York Moors, where no trace of the Chalk remains. Uplift and denudation were on a smaller scale further to the south, where, along with any Tertiary beds that might once have existed, about 100 metres of the Chalk were removed from the Wolds around Driffield.

Rivers changing course

It is difficult to recognise the earliest stages of landscape evolution, and there is debate over the little evidence that might be extracted from the modern landforms. The Chalk originally extended far to the north and west of its present outcrop. Subsequently there appears to have been uplift of a Pennine axis while the North Sea was subsiding, thereby creating a relatively uniform surface sloping gently towards the east – across the entire area now occupied by Yorkshire's Moors and Wolds. On this, trunk rivers drained eastwards. There is no single date when all this began, but the pattern was set sometime more than ten million years ago.

From beginnings in Wensleydale, it would appear likely that an ancestral River Ure continued eastwards to initiate a valley that would eventually become the Vale of Pickering, before it continued to the sea near Filey; it was probably joined by an ancestor of the River Nidd. Further south the River Aire was likely joined by the River Wharfe somewhere along its route to the Humber Estuary. In the north, the ancient River Swale probably joined the River Tees. There is a school of thought that maintains the Swale headed up the line of today's River Leven and then started to excavate an early stage of Eskdale out towards the sea

The major rivers of eastern Yorkshire. On the left: as they were before about ten million years ago, draining off the Pennines before the Vale of York and the eastern uplands were fully established. On the right: as they are today, with drainage converging on the Vale of York. The current outcrop of weaker mudstones is shown in pale yellow.

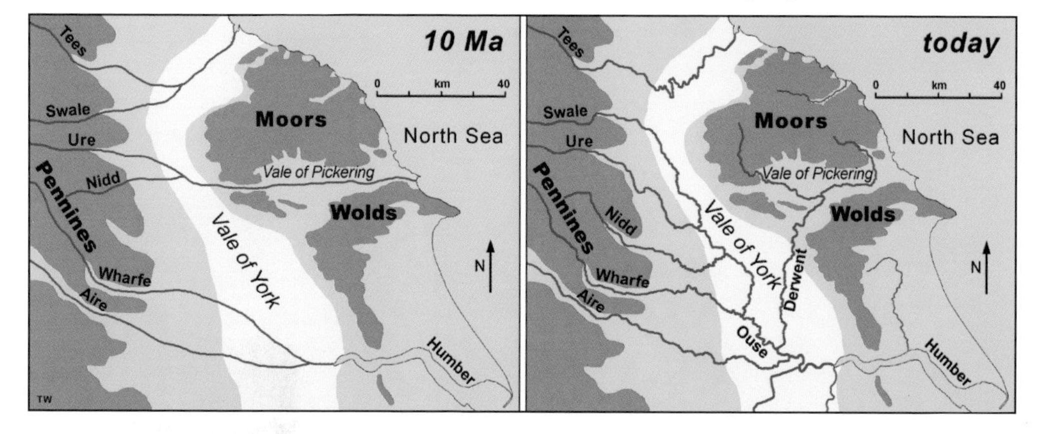

at Whitby, which neatly explains the low col between these two river basins; but there are others who consider that the Esk could have created its valley on its own.

Millions of years of slow surface lowering saw the removal of most of the Chalk, so that the pattern of major rivers was superimposed on to the more variable geology beneath – at which point the weakness of the Mercia Mudstone was revealed, and lowland developed over its entire outcrop, which happened to cut across the suite of eastbound rivers. Drainage always takes the easy line, so flood events across this new lowland allowed all those Pennine rivers to overflow and head south, thereby initiating the River Ouse. Meanwhile, drainage from the North York Moors and the Vale of Pickering continued to head eastwards – until abruptly changed by another big event. The River Derwent now flows westwards through the Vale of Pickering, reversing the pattern of its long-gone ancestor, but that grand turnaround took place millions of years later, when Ice Age glaciers intervened (*see* Chapter 7).

Outlines of the major components of the modern landscape were therefore established during those earlier times. The Vale of York, the broad outline of the North York Moors, and

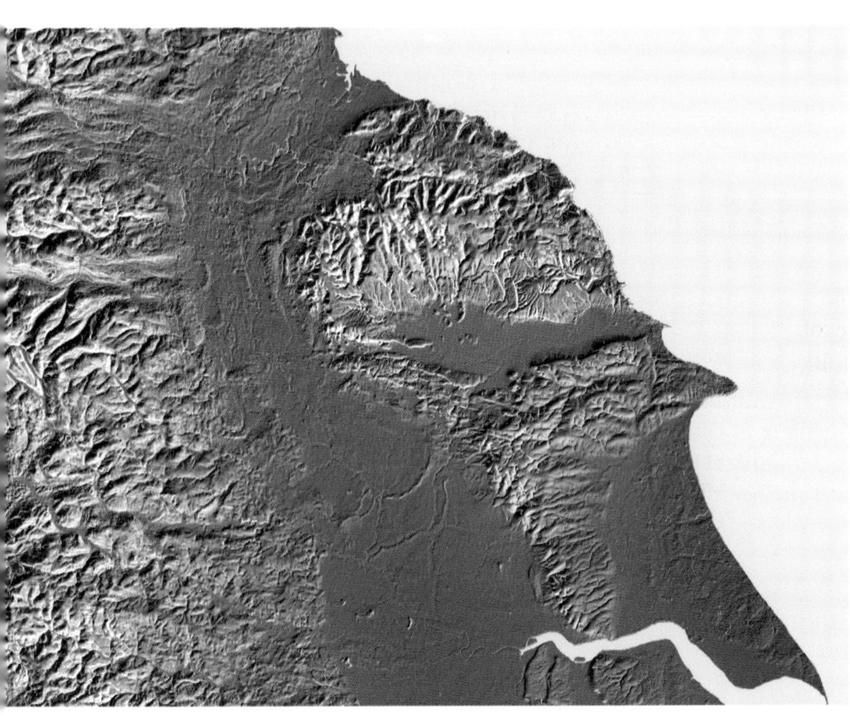

LEFT ***A digital terrain model of the eastern part of Yorkshire. Escarpments of the Moors and the Wolds both show with their northern and western parts higher and more dissected, above gentle dip slopes towards the south or east. The Vales of York and Pickering both show as uniform, flat grey, broken only by low hills of glacial till, most notably along the York Moraine.***

BELOW ***The upper end of Farndale, seen from Blakey Ridge; the valley has been cut down into the Redcar Mudstone, leaving the moors of Ravenscar sandstones high on each side.***

Great Fryup Dale in winter, looking to the southwest with the high Moors beyond. Eskdale is in the lower right, with the River Esk meandering between lines of dark trees and the railway having only one slight curve.

an ancestral version of the Vale of Pickering could well have been identifiable more than ten million years ago. Superimposed on to this broad pattern, the valleys, the hills, all the details and even the general line of most of the coast have emerged during surface denudation within the last few million years at most.

A major event in the evolution of the Moors, Wolds, Hills and Vales in Yorkshire was the Anglian glaciation, when ice sheets sourced on the high ground of Scandinavia and Scotland covered the entire region, and extended as far south as London. This, the most extensive ice cover ever in Britain, was lost around 425,000 years ago when world climates warmed into an interglacial stage. There were numerous Ice Ages prior to the Anglian, but all details of their landforms were removed or masked by the Anglian glaciers.

However, landform evolution is slow indeed when typical rates of surface lowering have been around 10 or 15 centimetres per thousand years. Major valleys can take a million years or so to reach perhaps 100 metres deep, and the hills are what is left behind. So the landscapes of the Moors and the Wolds have origins and ancestors that predate that Anglian glaciation.

For most of the last million years Yorkshire has been clear of ice, so most of the erosion and landscape development has been by water, in streams and rivers. Glaciers can be powerful erosive machines, leaving some distinctive landforms, but their overall role has been smaller than that of the rivers. For the last 425,000 years the high ground of the Moors and Wolds has been shaped by rivers. It was almost surrounded by ice sheets during the last Ice Age, the Devensian, which ended around 15,000 years ago, but that is another story, which applies to the low ground of eastern Yorkshire, and is told in Chapter 7.

Duck Bridge across the River Esk near Danby. The 14th-century packhorse bridge was high enough to clear any floods; the modern road crosses on a 'flat bridge' that has no parapets, so that floodwaters can pass over it with minimal obstruction.

High moorland, with the heather all brown, on the Ravenscar sandstones beside the main road above Newtondale.

Sandstone moors

Even though not entirely formed of sandstone, the North York Moors and the Cleveland Hills present a landscape that is draped with sandy soils and broken by sandstone scars. These thin, infertile soils support the classic moorland, which is effectively heathland with few trees, thereby offering the great panoramas of open space, except where conifer plantations extract some value from the land. Slow drainage on the almost flat Moors allows plants to rot in place and form patches of thin blanket bog. The natural vegetation is a mixture of heathers, bilberry and bracken along with various mosses and reedy sedges in place of fescue grasses. This creates a haven for wildlife, notably the grouse that thrive on young heather. Consequently many areas of the Moors are managed for the birds' benefit, with the side effect of ensuring the magnificent blaze of purple in late summer that is a hallmark of the North York Moors.

Across most of the high Moors and the Cleveland Hills, outcrops are dominated by the Ravenscar sandstones. By their quartz-rich composition and resistance to weathering, these define the nature of the moorland, but there is considerable variation within their rock types.

The north-facing escarpment of the Cleveland Hills, looking east from Battersby.

They include the resistant Moor Grit: this is a notably strong sandstone but is too thin, at around ten metres, to create more than a barely recognisable scarp along its zigzag outcrop across Spaunton Moor. Some Ravenscar beds have a grain size fine enough to be called siltstones, and there are also beds of mudstone that weather more easily. *En masse* they form the high ground of the Moors, but they do not create sharp escarpments that could match those formed by the even stronger Corallian beds.

The main rivers across the Moors drain southwards off the Cleveland Anticline where they were originally established on gentle dip slopes when that structure was first exposed. No longer do they stay atop the Ravenscar sandstones, since the major valleys – notably Rosedale, Farndale and upper Ryedale – have been cut deep into the mudstones of the Lower Jurassic. Once into the weaker mudstones, those valleys have widened out between retreating rims of stronger sandstones. Eskdale has a complex history that is not yet fully understood; it appears to have been superimposed on to a geological structure, though it, too, has widened out after cuttimg down into the weaker mudstones that underlie the sandstones of its flanks. Its southern tributaries, of Glaisdale and the Fryups, almost mirror the larger dales to the south.

Neither the Moors nor the Cleveland Hills have any lakes of significant size, and both are also short in man-made reservoirs. The dominant sandstones are a little too permeable for ideal reservoir sites, and most are too weak for the foundations of masonry or concrete dams. The two reservoirs on the northern flank of the Moors, at Lockwood Bank and Scaling, are each retained by low earth dams on areas of glacial till that overlie mixed sequences of sandstones and mudstones within the Ravenscar Group. The absence of natural lakes reflects the lack of recent glaciation, because it is valley glaciers that account for most lake sites, either by over-deepening fluvial valleys or leaving moraine dams within them – and there were never many of those in eastern Yorkshire.

Waterfalls add to the simple visual appeal of a landscape. However, in fluvial terrain that has not been recently glaciated, they largely depend on a stream or river cascading over a lip of strong rock to fall into a plunge pool in weak rock. Both Moors and Wolds lack contrast in the strengths of their various rock units, so waterfalls are not common in the region. The tallest is Mallyan Spout, with its stream falling twenty metres down a wall of Saltwick sandstone into the ravine of Wath Beck, west of Goathland.

The great northern edge of the Moors is created by the Cleveland Hills, which rise above the lowlands of the Tees Valley, to an altitude of 454 metres on Urra Moor, south of Great Ayton. Their steep, broken scarp faces are shrouded by woodland where the Ravenscar sandstones are not strong enough to stand out as bare crags; rock scars on stronger beds of sandstone, have probably all been modified by quarrying for building stone in the past. Highcliff Nab, high above Guisborough and visible from afar, is an impressive crag now popular with rock climbers, but much of its profile may well date from when it was cut back into a wooded hillside to provide stone for building Guisborough Priory.

One notable site where the rock is broken by nature and not by quarrymen is the Wainstones, on the crest of the hills south of Stokesley. This

A climber on the Wainstones, a tor of Ravenscar sandstone on the crest of the Cleveland Hills.

jumble of sandstone blocks appears to be a crumbled version of a tor that originated as a result of subsurface weathering picking out the fractures while leaving unscathed the larger intervening blocks. Surface lowering saw the subsequent removal of the smaller rock debris.

Roseberry Topping is a landmark crag, visible from afar. A tiny outlier of strong sandstone is an erosional remnant, separated from the plateau to its east, though even its summit may have been slightly modified by the removal of good stone for local building. The sandstone is the basal bed of the Ravenscar succession, so it sits atop a great cone of Whitby and Redcar mudstones, with the Cleveland Ironstone at about mid-height. The distinctive profile of the hill owes much to a great landslide on its southwestern face, and also to past mining of ironstone, jet and alum shale from the gentler slopes all around the rocky summit crag.

Roseberry Topping before and after it was re-profiled by the landslide in 1912.

South of the Moors, the Ravenscar Group also forms the dominant southern half of the double escarpment of the Howardian Hills, where the locally greater proportion of mudstones supports farmland and woodland in place of moorland. Castle Howard lies in the dip between the two escarpments, partly along the outcrop of the Oxford Clay, though faulting complicates the local geology and dictates a more broken and confusing landscape.

Roseberry Topping with its distinctive broken cap of Ravenscar sandstone above ramparts of Whitby Mudstone and Cleveland Ironstone.

Landslides on the moors

Interbedding of strong sandstones and weak mudstones creates ideal conditions for landslides, only ameliorated in the North York Moors by the low dips and relatively gentle slopes. The greatest numbers of slope failures are along the coast, simply because wave action and coast erosion generally achieve change more rapidly than fluvial erosion inland. Perhaps Yorkshire's best known recent landslide was the one that destroyed the Holbeck Hall Hotel at Scarborough; this was a classic rotational landslide developed largely within the thick glacial till (*see* page 94).

Inland landslides tend to be smaller and less frequent. Among the largest recorded events was the collapse of a large chunk of Roseberry Topping in May 1912. This occurred soon after a period of heavy rainfall, and it is likely that a rotational landslide developed over a curved slip surface within the steep slope of weak, saturated mudstone, driven in part by the load of the sandstone cap. The failure left an almost vertical head scar facing southwest above a chaos of massive sandstone blocks, with a wide fan of rock debris extending for 200 metres down the slope below. At the time, the collapse was blamed on the ironstone mining that was then active beneath almost the entire hill. Indeed, the mining may have weakened the hillside by opening up fractures as the roof rocks settled after being undermined. It is typical of landslides to have multiple causative factors, but it is usually water input that is the critical one.

The same applies to a great rockfall in 1755 that left the clean face of calcareous grit known as Whitestone Cliff, just north of Sutton Bank. Heavy rainfall could well have caused some movement and weakening, but collapse of the cliff face occurred late in March, during what is often known as the landslide season when the spring thaw releases rock from the grip of winter ice. Whatever caused it, Whitestone Cliff appears to have become more impressive by having its face so neatly trimmed.

While landslide events are not common, large numbers of prehistoric failures can be recognised by the deformed ground that shows up with an uneven or lumpy surface. Within the Moors, the prime landslide sites are the extensive valley-side outcrops of the Whitby Mudstone, especially where they slope steeply beneath a cap of resistant Dogger sandstone. Though not currently moving, many of these ancient landslides are prone to re-activation if saturated by unusually heavy rainfall events, or when their drainage has been disturbed by man's activities.

South of the Eskdale village of Ainthorpe, some of the plant cover was cleared from an area known as The Coombs during the 1980s. This accelerated the natural runoff, so a ditch was excavated to carry floodwater away from the channel that drained past Coombs Farm.

The face of Whitestone Cliff, near Sutton Bank, standing high above large blocks that fell away in the great rockfall of 1755.

RIGHT The graben that developed at the back of the Ainthorpe landslide, when ground on the left moved away downslope, so that the wedge in the centre dropped between small faults.

BELOW The minor road that was cut when the landslide near Ainthorpe was re-activated by a major rain storm in 1993.

However, the ditch's outfall was at the head of a landslide that was not then recognised, even though it had last moved as recently as 1929. A storm event in 1993 therefore reactivated the landslide, and broke the road between village and farm. It was a classic landslide entirely within the Whitby Mudstone. A block more than 100 metres across slipped just a few metres down the hillside, and a wedge of rock behind it dropped into the widening head-scar. Known as a graben, this down-faulted block had existed beforehand, but it dropped by another metre in 1993, taking the road with it. The road has since been repaired, albeit with a gentle dip in its line, and the graben can still be recognised as a bracken-strewn rift valley.

The Ainthorpe landslide is just one variant among the many slope failures that have occurred in the Whitby Mudstone. They have all contributed to the broken ground that distinguishes much of the Mudstone outcrop. Slope failures continue to occur. Each event may be dramatic or even destructive, but landslides are just one end of a spectrum of processes involved in slope erosion and surface denudation. Among recent events, two small rockfalls occurred within the town of Whitby,

The dark northern flank of Arden Great Moor where the Lower Calcareous Grit forms the northern tip of the Hambleton Hills.

both following intervals of heavy rainfall. In 2012 a slice of Ravenscar mudstones and glacial till slipped and destroyed five houses overlooking the inner harbour from below the old abbey. Then in 2013 a chunk of Ravenscar sandstones and mudstones fell away from the East Cliff above the outer harbour, with debris stopping just short of the houses below.

There will undoubtedly be more landslides in the future, particularly where weak and strong rocks are interbedded at outcrops along much of the coastal strip and on most of the valley sides within the North York Moors.

The step in the Moors landscape east of Saltergate, formed by the scarp face of the Lower Calcareous Grit along the northern rim of the Tabular Hills.

Hills of the Corallian

With roughly equal proportions of strong limestone and even stronger grit, and totalling somewhere over a hundred metres thick, the Corallian succession is a significant unit that defines its own landforms. Dipping at only a few degrees, it forms the broad plateaux of the Tabular and Hambleton Hills across the southern edge of the Moors, as well as the northern escarpment of the Howardian Hills. The conspicuous escarpment that zigzags along the northern fringe of the Tabular Hills is a splendid feature of the landscape, so clearly seen where the main road from Whitby to Pickering rises sharply on to the Lower Calcareous Grit at Salter Gate. Limestone and grit vie to form the larger area of the Tabular and Hambleton plateaus. Significantly the grit

The isolated tor of calcareous grit known as the Pepperpot, the narrowest of the Bridestones that stand on the Tabular Hills.

The Hole of Horcum, owing its shape largely to spring sapping at the base of the Calcareous Grit that forms the surrounding Tabular Hills.

is calcareous, so outcrops of both rocks are mantled in fertile lime-rich soils, and these Corallian hills support productive farmland, in marked contrast to the unproductive moors.

Narrow, steep-sided valleys dissect the Tabular Hills and carry the Moors drainage southwards, following lines established on bygone surfaces, dipping down from the Cleveland Anticline towards the structural low of the Vale of Pickering. Sections of the drainage loop underground beneath the outcrops of the limestones (*see* page 54), but the valleys continue unbroken between walls of grit. Even in the limestone they have been incised by floodwaters that overflow from constricted sinks, and also by meltwater when the ground was frozen during the various Ice Ages.

Despite the strength of the Corallian rocks, there are few bare cliffs or rocky crags. Every rule has its exceptions, and the Bridestones are lovely rock tors near the crest of the Tabular Hills, on the western edge of Dalby Forest. Formed of Lower Calcareous Grit (actually, the Passage Beds that are transitional up into the Hambleton Oolite), each of the half-dozen rock towers stands about five metres tall and has been weathered to a jagged profile that picks out strong beds and weak fractures. They are classic tors in that they are residuals left behind because they were the largest unbroken blocks surviving within a bed of sandstone riven by fractures

with variable spacing. They owe much to subsurface weathering that saw disintegration of the intervening fractured rock before it was rapidly eroded away when exposed by surface lowering. Elsewhere, many comparable bits of isolated rock architecture have been shown to be quarry remnants, but the Bridestones are handiworks of nature.

The pretty village of Hutton-le-Hole straggles down both sides of Hutton Beck where it drains down-dip on the sheltered outcrop of Osgodby sandstones just north of the grit escarpment of the Tabular Hills.

Roulston Scar forms the edge of a plateau of Lower Calcareous Grit, which is part of the Hambleton Hills, and a tiny outlier of the same grit survives on the outlying Hood Hill.

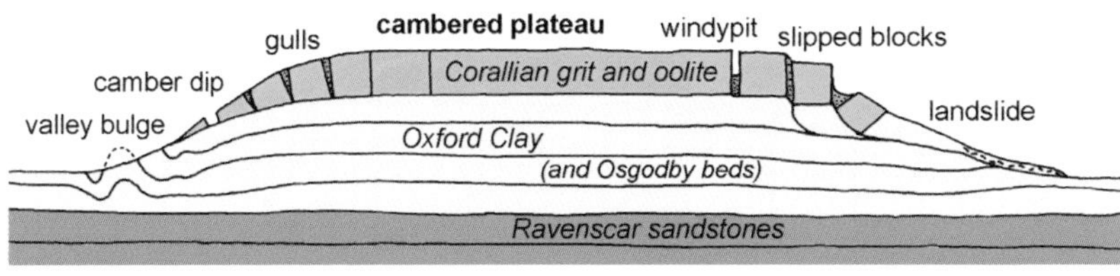

Profile through a cambered plateau with the various features of edge failure (not to scale).

Hambleton Oolite in Kepwick Quarry. In this view, the regional dip is towards the left, but the beds dip right due to camber folding over the soft Oxford Clay; the main fractures are gulls, opened up by the blocks of limestone sliding towards the right.

West of Dalby Forest, the main road across the Moors skirts the rim of the Hole of Horcum. This dramatic and entirely natural bowl is a kilometre in diameter, with a steep rim of Calcareous Grit 50 metres tall overlooking an almost flat floor in Oxford Clay and the underlying soft Osgodby sandstone. It appears to have been formed by headward sapping at a ring of springs ranged along the boundary where permeable grit sits on impermeable clay. Though the bowl may conjure up comparisons with some glacial landforms, and sluggish ice must have covered the site during the Anglian glaciation, there is no evidence that ice had any significant role in excavating the Hole of Horcum. It is a fluvial landform, albeit an unusual and rather splendid one.

Though generally lacking in scars and crags, the Corallian rocks are dramatically exposed at both ends of the escarpment. At its eastern end the Lower Calcareous Grit breaks out from beneath its cover of glacial till, to form the vertical upper half of Newbiggin Cliff, and then lowers in height eastwards to form the narrow base of Filey Brigg. However, it is Birdsall Grit that forms the outer part of the Brigg extending far out to sea at low tide.

At the western end of the outcrop, Whitestone Cliff and Roulston Scar stand clear, respectively north and south of the well-known Sutton Bank on the road from Thirsk to Scarborough. Both are vertical walls of

Castle Howard stands between two lakes that are both artificial in the heart of the Howardian Hills.

Lower Calcareous Grit that stand tall above lower ramparts of Ravenscar sandstones extensively disturbed by small-scale landslips. It appears that the stability of these grit cliffs might owe much to the local absence of the Oxford Clay that normally underlies the Calcareous Grit. That clay is so soft that, along valley sides, it is commonly squeezed out from beneath the edges of the stronger grit above it. Then the edges of the grit subside (this is known as camber folding, after the road-like profile it creates across a plateau); the effect is widespread along the sides of Ryedale (*see* page 62), where the cambered grit is fractured simply because it is brittle. Without such deformation above Oxford Clay, the grit at Sutton Bank is less broken and all the better to form its grand cliffs. Just eight kilometres to the north, the Grit and the overlying Hambleton Oolite (exposed in the old Kepwick Quarry) are conspicuously cambered and heavily fractured above the thick Oxford Clay that has been squeezed out from beneath the escarpment edge.

Chalk wolds

By no means can chalk be described as a strong rock, but it is very good at forming high ground that overlooks lowland on neighbouring outcrops of other rocks. Such an apparent anomaly can be explained by the rather lovely concept that 'chalk devours its own agents of erosion'. This refers to rainfall sinking straight into the porous and very permeable chalk, so that streams and surface water are rarities. Consequently, the soft and weak chalk is topographically resistant, simply because there is little to erode it, and it forms significant hills by default. The Yorkshire Wolds are a prime example, with their distinctive landscapes of rolling hills between dry valleys (*see* page 58). By the same means, the chalk coastline is distinguished by great white cliffs that have no streams to erode their crests, while waves eat into their bases.

The rounded profiles of dry valleys that are typical of the chalk karst on the Yorkshire Wolds, at the head of Thixendale, high on the dip slope west of Driffield.

Grain crops occupy nearly all the land within the Great Wolds Valley west of Weaverthorpe.

The Wolds form a classic escarpment, albeit one that is gently graded and is bent in the middle. Their main arm extends westwards from Flamborough Head, rising steadily to a high point, a modest 246 metres above sea level, on Bishop Wilton Wold, some seven kilometres north of Pocklington. With no scars or crags, the scarp face of the Wolds escarpment presents a conspicuous barrier along the southern margin of the Vale of Pickering. In contrast, the Wolds' dip slope, deeply scored by many twisting branches of the dry valleys, descends imperceptibly towards the southeast until it is lost into the low country of Holderness. The Wolds' southern arm is more of the same, forming a lower and narrower belt that overlooks the Vale of York until it ends where the chalk outcrop is breached by the Humber Gap.

Devensian ice never crossed the crest of the Wolds escarpment except for about seven kilometres of its eastern extremity in from Flamborough Head. Once over that shoulder, the North Sea Ice Sheet spread a little westwards and smothered the lower part of the chalk dip slope that is now lost beneath the glacial till of Holderness. However, Anglian ice had previously over-run the entire chalk outcrop, though it is difficult to see any impact that remains in the modern landscape. The last remnants of Anglian till, or indeed any other pre-Devensian till, were probably washed away by snowmelt when surface streams were at their maximum over frozen ground in Devensian times.

Across the heart of the chalk country, the Great Wolds Valley extends from Duggleby and West Lutton eastwards to Burton Fleming, where its outlet turns south to Rudston and then out to the sea at Bridlington. For its main part, it is around three kilometres wide with gentle flanks rising 50 metres or so to its bounding ridges. Clearly a very old feature, it was almost certainly over-run by Anglian ice, which would have flowed across, and not along, the pre-existing fluvial valley. There is, however, some mystery about the Great Wolds Valley. Unlike any other Wolds valley, its alignment towards the east across the main northern arm of the Wolds runs across the dip slope, instead of roughly down the dip. This could be due to early guidance by a fracture zone within the chalk, or possibly by some subglacial drainage pattern, but neither seems to be an adequate explanation. Its broad track across the Wolds now holds a swathe of productive farmland, and is traversed by the Gypsey Race (*see* page 60), a stream that is yet another rather unusual feature of this chalk landscape.

Clay vales

Whether defined as mudstone, shale or clay, these fine-grained sedimentary rocks, which form much of the Mesozoic succession in eastern Yorkshire, are easily eroded and therefore define the low ground within the landscape. Shale is a laminated variant of mudstone, and both of these weather into a softer material that is better described as clay. The inherent weakness of all three is then picked out by selective fluvial erosion whereby rivers and streams etch the detail into the terrain.

Hardly a detail, the Vale of York is a broad swathe of lowland stretching from between the Tees and Humber valleys, along the western side of both the Moors and the Wolds. It was initiated on the adjacent outcrops of the Triassic Mercia Mudstone and the Lower Jurassic Redcar Mudstone. Those two beds still underlie the eastern half of the Vale, but the western half has been cut down into the Sherwood Sandstone. This is also a weak rock, very different from the stronger sandstones of the Moors or the Pennines, so it has been easily eroded by the Vale's suite of converging rivers, and also by significant ice streams during glacial interludes of the Quaternary. Unfortunately, surface lowering within the Vale, followed by deposition of alluvium, has left the River Ouse at York only a few metres above sea level with more than 50 km still to flow to reach the open water in the Humber. The effect of this is extensive flooding within the Vale soon after major rainfall events in its Pennine catchment.

Waterside inns and cafés along the River Ouse where it passes the centre of York.

Yet another flood event at York, where the River Ouse has to flow through its gap in the York moraine.

The Vale of York, seen from the edge of the Wolds.

In stratigraphic sequence, the next major clay band is the Oxford Clay. Though a major feature in its type locality in southern England, its thickness is reduced to 30 metres or less within eastern Yorkshire. But it punches above its weight with its influence on the local topography. Eroded down to a narrow band of lower ground, it defines the clean break between the North York Moors and the Tabular Hills, undermining the latter to create a long but low escarpment as the northern rim. The impact of the clay has been even greater in Ryedale; this is a lovely box valley with steep wooded flanks on each side of an almost flat floor that is the setting for the spectacular remains of Rievaulx Abbey. Unaided by any bygone glaciers, the River Rye has cut down through the strong Corallian grit and limestone, to then expand laterally once into the soft Oxford Clay beneath. Subsequent undercutting has seen retreat of the strong rock walls, turning an original narrow canyon into the broad box valley of today. With similar effect on a smaller scale, Oxford Clay has defined the spring line that caused expansion of the Hole of Horcum within the Tabular Hills, though there the aquifer feeding the springs is Calcareous Grit alone.

A view south across the Vale of Pickering, with the Wolds along the skyline.

Third of the big clay units is the Kimmeridge Clay. Lying between the Jurassic sandstones and the Cretaceous chalk, this accounts for the wide floor of the Vale of Pickering. Quite possibly originally excavated by an ancestral River Ure, it could equally well have been developed by local rivers eroding the inherent weakness of the clay outcrop. Either way, the Vale was considerably modified by Quaternary events when it was occupied by a great lake (*see* page 80). The western end of the Vale is almost closed off by the structural rise of the Corallian beds that form the Howardian Hills. However, that closure is incomplete, as the Ampleforth rift valley is a conspicuous fault-guided depression separating the Howardian and Hambleton hills. The breakout into the Vale of York is known as the Coxswold Gap, which could perhaps owe its entire existence to the structural geology, with no need to invoke any role by a long-gone River Ure on its way into the Vale of Pickering.

Glacial till is weak, unconsolidated, unsorted sediment that commonly has a significant clay component, to the extent that it was called boulder clay in the past. But it is not a fourth great clay unit defining lowlands in East Yorkshire. Conversely, it forms small areas of higher ground at many places, simply because it was deposited by Ice Age glaciers so recently that it has not yet been whittled down by post-glacial erosion. Holderness is a lowland, but its glacial till is only a veneer. Holderness lies low largely because its chalk bedrock descends into the tectonic basin occupied by the North Sea. Indeed, the eastern parts of Holderness only exist because the glacial till forms dry land where the top of bedrock is below sea level.

The clays of East Yorkshire form neither Moors nor Wolds, but they underlie all the low ground that surrounds those two splendid areas of upland. In many regions away from Yorkshire, rivers and glaciers more or less ignored the geology to create their own landforms. But the contrasts between strong and weak rocks within the sedimentary sequence of eastern Yorkshire mean that the geology defines the nature of the landscape on a rather spectacular scale.

CHAPTER 6

Karst and Caves

A karst landscape is one with its drainage underground. This normally means that it has cave systems and is recognisable by its distinctive landforms of closed depressions, sinkholes and dry valleys. Karst is therefore formed on rock that is dissolved in rainwater, which means largely limestone or gypsum. And because soil is not created by weathering of soluble materials, most karst is distinguished by its landscapes of bare white rock.

Yorkshire's Moors and Wolds have no karst landscapes that conform to the standard definition, but they do include two terrains that have their own varieties of karst. The Tabular Hills have cave drainage but lack most of the diagnostic karst landforms because they are formed of a mixture of limestone and sandstone. And the Yorkshire Wolds are a special variety known as chalk karst.

Karst in the Tabular Hills

Lying all along the southern edge of the Moors, the Tabular Hills are formed of two oolitic limestones sandwiched between three calcareous sandstones, collectively known as the Corallian Group (*see* page 21). The limestones form, or at least cap, about half of the low plateaus that account for their name as the Tabular Hills. However, these are barely recognisable as karst, because they are blanketed by thick soils, derived in part from weathering of the sandstones that have since been removed by erosion.

In contrast, the floors of the many valleys through the Tabular Hills do bear the hallmark features of karst. All the streams draining south from the Moors towards the Vale of Pickering, lose at least some of their water

A dry valley cut into the Hambleton limestone on the dip slope of the eastern Tabular Hills north of Snainton.

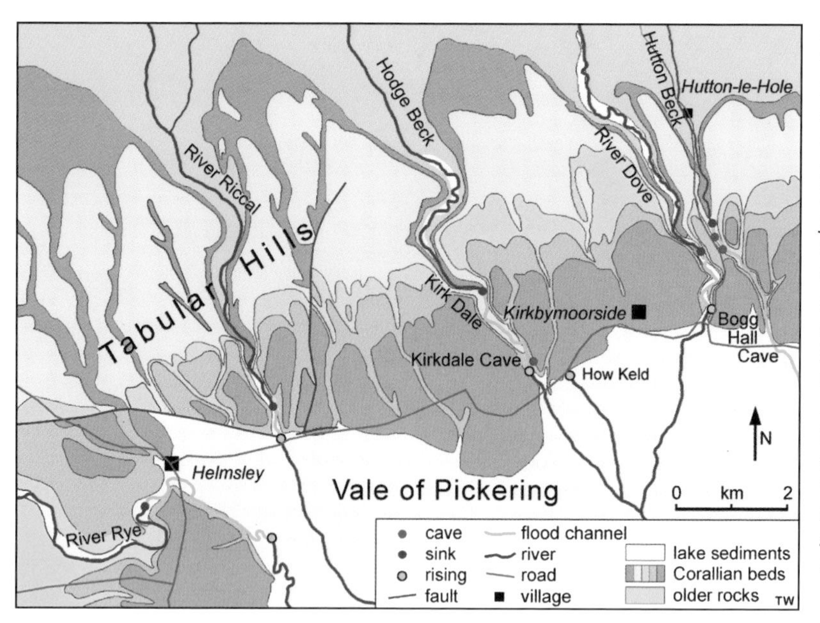

Geology of the Tabular Hills, with the two limestones in blues and the calcareous grits in yellow-browns** (see **Table on page 21 for the Corallian succession), showing the rivers that drain underground across outcrops of the limestone. Outcrops of Kimmeridge Clay are not marked; they form small exposures in the Vale of Pickering and thin caps on some hilltops of the Calcareous Grit.

underground where they cross the outcrops of the limestones. In dry weather the sinks can be recognised where the water disappears into the streambed alluvium or bedrock fissures, but there are no gaping sinkholes. Water then emerges at springs along the edge of the Vale of Pickering; most of these are quiet pools with water flowing out of them, but four of them produce enough clear water to feed trout farms. Between sinks and risings, the streambeds are dry in dry weather, but carry plenty of floodwater after significant rainfalls.

Flow in the River Rye sinks underground where it passes through Duncombe Park, and reappears two kilometres to the east where it feeds a fish farm, but both sink and rising are unexciting sites in the channels' alluvium. A large spring at the head of a stream or river is commonly named a *keld*. Howe Keld, southwest of Kirkbymoorside, and Keldhead, on the western edge of Pickering, are two notable sites along the edge of the Vale of Pickering. They are the natural risings for underground streams that sink into the Corallian limestones where they flow off the Moors and into the Tabular Hills; but the flows from both are now reduced because water is pumped from boreholes drilled down to the aquifer. Further east, the upper River Derwent now has part of its flow diverted into a man-made short-cut out to the sea, but its natural course through the narrow Forge Valley still loses water underground across the limestone outcrop. Previously this reappeared at springs in the Vale of Pickering, but most of it now flows to boreholes that abstract water near Scarborough, thereby distorting drainage patterns within the limestone aquifer.

Keldhead, a pool that is underlain by a powerful rising on the western edge of Pickering; it yields underground drainage from the karst of the Tabular Hills into the head of Costa Beck.

All these underground flows are through fissure systems within the limestone, and many of these fissures have been enlarged by dissolution of their walls so that they now form cave passages. But little is known about them. Most sinks and risings are impenetrable, and many of the stream caves are probably too small to enter anyway. Bogg Hall Cave, the rising at the head of the River Dove east of Kirkbymoorside, was one of the few with any accessible passage, until longer caves were found at the sinks that feed it.

A cascade of calcite flowstone down the wall of a tall section of the main stream passage in Excalibur Pot.

Caves beneath Hutton Beck

In 2007, local cavers cleared the rubble and sediment out of choked fissures in the bedrock along the channel of Hutton Beck, which is left dry when the beck's fair-weather flow is all lost into its Top Sink. Eventually they opened up Excalibur Pot and Jenga Pot, both of which descend into a cave system where more than four kilometres of passages are now known. Most of these are not large, and nearly all of them are liberally coated with sticky brown mud. But there are a few sections of walking-size passage along various stream routes, and some of these are adorned with stalactites and flowstone. The cave streamways carry water from various sinks along Hutton Beck. The River Dove also sinks underground, and joins the Hutton Beck water beyond the current limits of exploration. All the cave streams coalesce before their water returns to daylight at Bogg Hall Cave, where only a short length of passage has been explored to date.

A tributary flow enters the main streamway of Excalibur Pot from a tiny bedding-plane passage.

The head of the River Dove, where the combined waters of the upper Dove and Hutton Beck re-appear from the Malton Oolite at Bogg Hall Cave, set into the southern edge of the Tabular Hills.

The sinks along both Hutton Beck and the River Dove, and all the known cave passages beneath them, are formed in the Hambleton Oolite, but their waters eventually emerge from Bogg Hall Cave in the overlying Malton Oolite. Somewhere in the unexplored section of the stream routes, the waters break upwards from one oolitic limestone to the other. Whether the intervening Middle Calcareous Grit is locally absent, is offset by an unknown fault, or is breached by open fissures, remains unknown; and it will contnue so until the breakthrough point is reached when more of the stream passages have been discovered and mapped.

The system of cave passages now known to lie beneath Hutton Beck has to be indicative of caves and stream routes that lie between sinks and risings in each of the valleys through the Tabular Hills. Many of these cave passages may be small and muddy, but there are surely more to be discovered within this little-known karst.

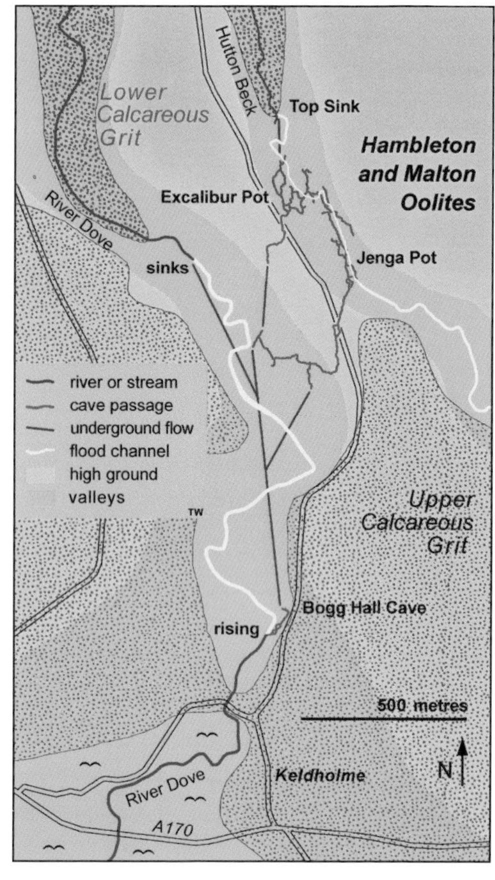

RIGHT ***Underground drainage and cave passages that are known below the River Dove and Hutton Beck.***

BELOW ***A low stream passage that has developed along a gently-dipping bedding plane in Jenga Pot.***

Kirkdale Cave

In the summer of 1821, a small quarry in the mouth of Kirk Dale, where it emerges from the southern edge of the Tabular Hills, broke into a fragment of old, dry cave passage. It is easily accessible, lying just above the ford where the old road crosses Hodge Beck. The cave follows the boundary between the Malton Oolite and the Coral Rag, and the quarrymen crushed the cave's walls and contents to produce their roadstone. Numerous bones within the cave were assumed to have been dumped there during some past epidemic, and were used to fill potholes in nearby roads. It was pure good fortune that local residents found some of these bones in the repaired roads, and traced their source to the quarry.

More of the cave sediments were then excavated by a reverend gentleman from Whitby. He recognised the bones as being much older, and described them as being washed into the cave by Noah's biblical deluge. Even when some were identified as being from rhinoceros and elephant, he had no problem in claiming that they had been carried to Yorkshire by the great flood surge from the Middle East.

Word of the finds eventually reached Reverend William Buckland at the University of Oxford. He travelled to Kirk Dale, recorded the bones of 23 animal species, and concluded that the cave had been a hyaena den. The low bedding-plane cave was not large enough to have been used by early mankind. Flowstone covering some of the sediment has since been dated to 121,000 years ago, so the bones are ascribed to the Ipswichian Interglacial. They include teeth of hippopotamus that are the furthest north ever discovered, and these indicate a contemporary climate rather warmer than that of today. The actual cave passages are even older, formed in some previous interglacial stage when they carried drainage between sink and rising, neither of which can now be identified; and that was well before they were left high and dry to provide a welcome den for roaming hyaenas.

The present entrance to Kirkdale Cave, created when the cave passage was breached by the quarry face that was being excavated into the Malton Oolite (below the cave) and the Coral Rag (above the cave).

The quarry has long ceased production, and no bones remain within the cave passages, of which only about ten metres were removed by the quarrymen. Kirkdale Cave will for ever be known for having yielded a magnificent assemblage of animal bones that provide a picture of conditions at a time between the glacial invasions of northern England.

Chalk karst of the Wolds

A special kind of karst landscape develops on the variety of limestone known as chalk. There are no bare crags because chalk is softer than other limestone, though the coastal cliffs are on impressive scales. There are almost no caves large enough to enter, because the entire rock is so porous that groundwater just diffuses through it. The chalk forms high ground, because it is barely eroded when rainwater sinks underground before it can gather into streams.

The gently undulating chalk hills are known, by a quirk of the English language, as downs, or downland; the term originates from the Celtic *dun*, meaning a hill. The Downs are features of southern England, and their northern cousins are known as Wolds.

ABOVE Sheepwalk grassland that is so typical of the chalk Wolds, here in Cottam Dale, north of Driffield.

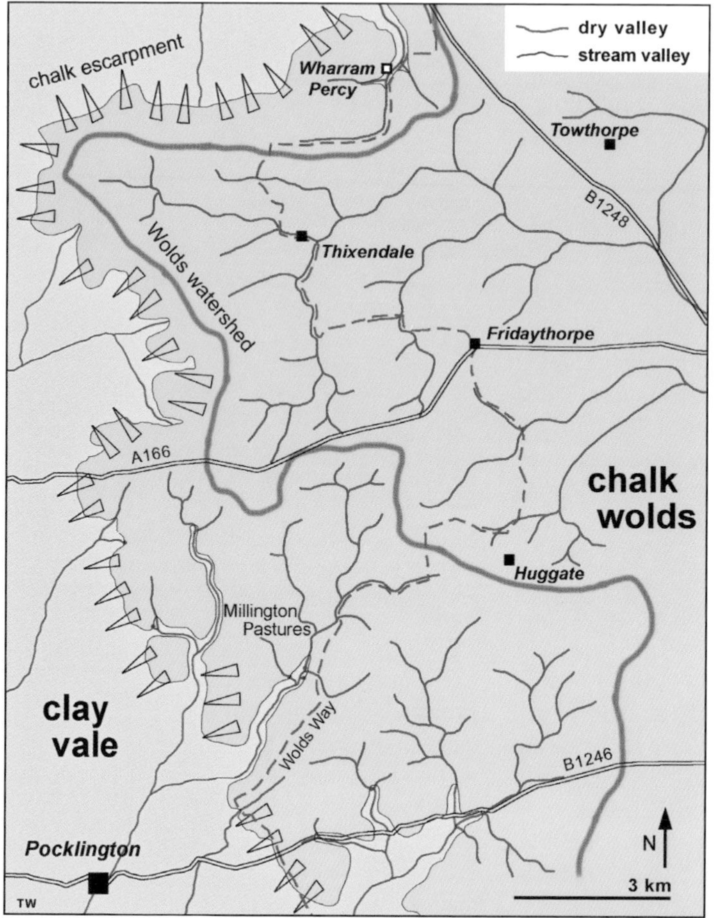

LEFT Dendritic systems of dry valleys in the northwestern corner of the Yorkshire Wolds; east of the sinuous watershed, these are aligned down the dip slope of the escarpment, whereas those to the west of the watershed are cut steeply back into the scarp face.

A wold is defined as a rolling upland largely devoid of tree cover, though the name derives from the Old English *wald*, meaning a forest and reflecting their woodland cover prior to clearance by the early farmers. Both downs and wolds are characterised by their cover of short, dense turf, which is often described as sheepwalk. Its grass is short because sheep nibble the tops while growth continues from beneath. Grazing sheep also nibble the tops off any other plants, shrubs or trees, all of which grow at their tips and are therefore cut short in their prime. Only in later years have most of the Yorkshire Wolds been developed for productive agriculture. Huge fields of grain now dominate the landscape, though steeper slopes along the deeper valleys remain as classic sheepwalk grassland.

Both agriculture and sheepwalk grassland benefit from the fine-grained soils that overlie the chalk. These are largely formed of loess, which is typical of periglacial terrains because it consists of silt that was blown by the wind from fresh glacial sediments. Periglacial means peripheral to the glaciers. Loess is widespread downwind from any area of unsorted glacial debris that is temporarily dried at the surface. However, it is normally mixed in with soil that is produced by *in situ* weathering; only on chalk and limestone does it dominate because these rocks go into solution without creating mineral soils. Generally around half a metre thick across the Yorkshire Wolds, loess contributes to the seasonal changing of landscape colours: in spring the green of young shoots, in summer the gold of mature grain crops, and then back to the soft brown of bare loessic soil.

Dry valleys are the ubiquitous landform of chalk karst. They are essentially periglacial features, excavated by surface streams and rivers that could not sink into the frozen ground. Ground that is frozen to depths of a hundred metres or more is known as permafrost. Only the top metre or so of ground thaws out each summer, and it refreezes during the autumn. Winters are cold and thick snow cover accumulates, followed by spring thaw with powerful streams fed by melting snow. Permafrost is widespread in arctic lands today, and was a feature of Britain during the Ice Ages. That was when fluvial erosion dominated and valleys were cut ever deeper.

West of Thixendale village, Water Dale has no permanent stream, as is typical of valleys in the chalk karst.

Throughout the Devensian cold stage, the Wolds were periglacial when they were almost surrounded by great ice sheets extending from the north. Devensian ice reached over the chalk high ground a few kilometres inland of Flamborough, and spread over the lower chalk outcrops that now underlie Holderness. Ice covered the entire area during the Anglian glaciation, though periglacial conditions pertained during phases at the start and end when the ice sheet was advancing and retreating. So the Wolds were clear of ice and underlain by permafrost through most of the Devensian, in parts of the Anglian, and during an unknown number of the other cold phases that distinguished the Quaternary (and some of which were marked by Ice Ages in other parts of the world). That gave plenty of time, in multiple phases, for streams and rivers to excavate the valleys that are now deep and dry within the Wolds. Meltwater from the snow packs did most of the work, cascading off the hills in the warmth of each spring season.

Deepest of the chalk dry valleys are those entrenched into the higher ground on both sides of the Wolds watershed. The splendid Millington Dale and its tributaries are entrenched by around 70 metres into the heavily dissected scarp face of the Downs escarpment, and the lovely valleys that converge on Thixendale are cut nearly as deeply into the dip slope out towards the east.

Rather different is the Great Wolds Valley, heading east across the gentle dip slope of the chalk. This broad vale appears to pre-date the Anglian glaciation, when ice from the north would have swept across it, thereby tending to flatten its profile, instead of deepening it, as would have been the case if ice had flowed along it. The great valley has the longest stream course within the Wolds, going under the delightful name of the Gypsey Race. The gypsey name implies some mysticism, related to the fact that it only flows intermittently. It is a classic chalk bourne, which only flows after wet weather raises the local water table to

One of the many beautiful dry valleys that are tributary to Millington Dale in the well-developed chalk karst on the scarp face of the Yorkshire Wolds northeast of Pocklington.

The dry channel of the Gypsey Race where it passes through the village of Burton Fleming; the bridge is necessary as long periods of very heavy rainfall can raise the water table within the chalk and thereby generate temporary surface flows down the bourne.

ground level, whereas in dry weather it ceases to flow above a declined water table. For much of the time the Gypsey Race plays ducks and drakes, rising from tiny springs, flowing to obscure sinks, and then re-appearing further down the valley. From Helperthorpe down to Weaverthorpe the channel is permanently dry, except for local run-off, but other sections are normally dry and then carry flows after wet weather. Once on to the glacial till near Boynton, the Race's flow is more regular, and it is culverted through parts of Bridlington before draining into the harbour.

The Great Wolds Valley has another claim to fame in that in 1795 it was the target of the largest meteorite known to have landed in Britain. A stony chondrite, it weighed in at 25 kg, and was seen to fall from the sky to make a crater a metre across; the site is now marked by a brick monument a kilometre southwest of Wold Newton.

Chalk springs and holy wells

Rainwater that sinks into the chalk has to emerge somewhere. In the past it has drained to numerous springs, many of which became the sites for early settlement. With a reliable water supply so important to life, many chalk springs were regarded as holy, and some became known as holy wells, even though they were natural springs and not dug wells. St Helen's Well is one such, in the narrow valley east of Market Weighton, and right on the base of the chalk. Along those western and northern margins of the Wolds, nearly every dry valley cut into the steep scarp face of the chalk escarpment gains a spring and a stream where it crosses the base of the chalk on to older clays. The now deserted medieval village of Wharram Percy was carefully sited in a sheltered valley just downstream of springs from the chalk; these still feed into the village's reservoir that stands on the Kimmeridge Clay.

Away from that narrow scarp face, most groundwater beneath the Wolds drains down-dip and down-slope towards the east and south. The Keld is a major rising from a tree-lined pool on the western edge of Driffield, and powerful springs around Beverley were just some of many more that fed tributaries of the River Hull.

The Mere, in the centre of Nafferton, is fed by a cluster of springs where clean and clear groundwater rises from the chalk where the Ipswichian cliff is buried beneath the glacial till.

Water in these larger springs emanates from the chalk and breaks up through the feather-edge of the glacial till, or rises through permeable patches within it, whereas others were simply overflows from the chalk where it becomes trapped beneath the till. However, large-scale abstraction of groundwater from deep wells has lowered the regional water table, with the result that many springs no longer flow, and the survivors are pale shadows of their past importance. Municipal supplies are now assured, but their resources are not infinite. A proportion of the chalk groundwater originally flowed further east to emerge from sea-floor outcrops. Excessive abstraction from the deep chalk wells means there is less water flowing seawards, and consequently flows have reversed and saltwater is flowing inland. This is known as saline intrusion, and is a major constraint on groundwater abstraction from the coastal chalk.

Despite so many ancient springs ceasing to flow, some do survive. Out on Holderness, holy wells in both the villages of Atwick and Great Hatfield are actually small springs. Their unceasing flow suggests that they are not fed from the deep and depleted chalk aquifer, but are sourced in permeable horizons at shallow depths within the glacial till. Neither is now used to supply its villagers, which is perhaps fortunate, because some of their localised and shallow drainage is probably derived from cattle pasture not very far away. Springs should not be regarded as gifts from the gods.

Windypits in the Hambleton Hills

Scattered along the edges of the Hambleton Hills, the windypits derive their name from their habit of emitting strong flows of warm air during winter. They are not karstic features but are rather magnificent landslip fissures that are something of a local speciality. The strong Corallian grits and limestones sit on top of weak and slippery Oxford Clay, and it is very easy for entire hillsides of Corallian rock to slide very slowly into the adjacent valley. This creates great vertical fissures, some of which are 50 metres deep and more than 100 metres long. They are typically no more than a few metres wide, and features of their walls match where they have simply been pulled apart. They do not appear as open fissures because their cover of soil and vegetation deforms to keep pace with the slowly creeping ground, and shallow blocks can also slide across any gaps. But little collapses do occur; these then provide the draughting entrances, and thereby reveal access into the great fissures that would otherwise remain unknown.

Windypits are scattered round the edges of the Hambleton Hills, with the largest along the rim of the southern end of Ryedale and its tributary valleys. The tectonic dip is only a few degrees; that is barely enough to generate sliding of the hillside blocks, and the local dips are likely helped by a little camber folding (*see* diagram on page 48). This occurs where soft underlying clay is squeezed out from beneath the edges of a plateau: the displaced clay causes the adjacent valley floor to bulge upwards, though river erosion removes the clay as fast

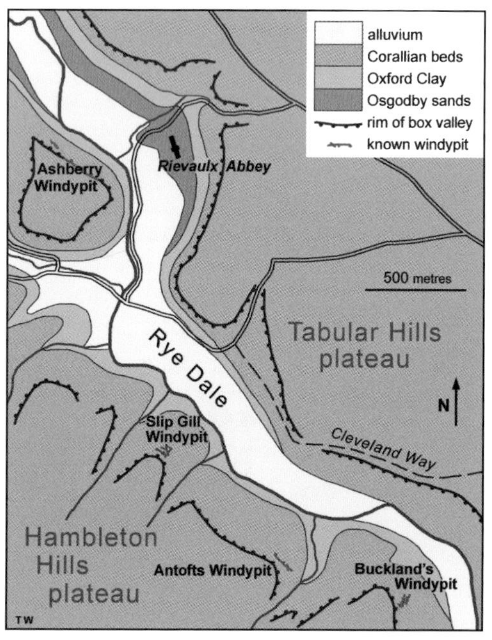

Windypits that have been explored along the southern flank of Rye Dale downstream of Rievaulx Abbey.

as it is uplifted. With the clay squeezed out, the edge of the plateau sags a little, and develops a camber, like a road profile. The bending also develops deep fractures, and these make landslip easy, with the dipping blocks sliding over the cambered top of the clay.

The windypits of Ryedale include some complex mazes of fissures between huge slipped blocks, but Antofts is probably the nearest to the archetypal windypit. A single fissure lies parallel to the hillside. More than 100 metres long, most of it is open to heights of more than 15 metres. It is little more than a metre wide, as that is the distance that the hillside block has slipped outwards. Its floor is a chaos of fallen blocks, and its roof, some 10 metres below ground level, is formed by overlying beds that have not been displaced as far. A collapse in a small crossing fissure provides the only way in.

There are clearly many more great open fissures inside the slopes of Ryedale, but each will remain unseen until some almost trivial collapse of the ground reveals an entrance. The windypits are not karstic features and they can never qualify as visitor attractions, except to enthusiastic cavers. But they are rather special features of the Hambleton Hills – geology in action, in its usual manner that is very slow and rarely visible.

The tall fissure that is the core of the windypit known as MSG Hole, west of Ryedale.

Ryedale, seen looking upstream from above Rievaulx Abbey, with the woodland hiding Ashberry Windypit within the steep slope on the left, and likely many more landslip fissures, as yet undiscovered, on both sides of the valley.

CHAPTER 7

Ice Age Glaciers and Lakes

Locals call it the Inland Ice – we call it the Greenland Icecap – and it has been there for a very long time. Made up of repeated winter snowfalls compressed into ice by the weight of successive falls, the lowest layers are the oldest and are only reached by deep boreholes. Layers can be dated by the unstable isotopes within the ice's impurities, and world mean temperatures can then be interpreted from the ratios of stable oxygen isotopes within air bubbles trapped within the same layers.

Hence we have a record of world climates that reaches back for thousands of years. Now combined with, and extended by, the same isotope ratios in ocean-floor shell debris, the isotope record defines the stages of climatic oscillations through the last few million years of the Quaternary period of geological time. Odd numbers are warm, even numbers are cold. We are now in Marine Isotope Stage 1 (MIS 1), which is also described as post-glacial, or as the Holocene.

Climatic fluctuations recognised as MIS 4 and MIS 2 encompassed the Devensian glaciation (its name derives from Deva, the Roman camp at Chester, where the glacial sediments are extensive). In formal terminology, the Devensian is a stage within the Quaternary period, though it includes a number of isotope stages; it is colloquially known as an 'Ice Age'. Devensian ice sheets reached their greatest extent between about 27,000 and 19,000 years ago. That event is known as the Last Glacial Maximum, after

Thick glacial till is exposed at Skipsea and all along the Holderness coast.

climate	*stages*	*MIS*	*years BP*	*impact on Moors and Wolds*
warm	Holocene	1	11,700	post-glacial details
cold	Loch Lomond	(2)	12,000	minor
warm	Windermere	(2)	15,000	minor
cold	main **Devensian** (Last Glacial Max)	2	29,000	glacier ice to west and east, ice-dammed lakes on Moors
warm/cold	early Devensian	3-4	116,000	uncertain
warm	Ipswichian	5e	128,000	Kirkdale Cave hyaena den
cold	Wolstonian	6-8	200,000	icefield on Tabular Hills
warm	Hoxnian	11	425,000	river erosion
cold	**Anglian**	12	480,000	total ice cover
warm	Cromerian	13		unknown details
and cold	earler glaciations	etc		major landscapes evolved

Stages and features of the Quaternary that are significant in eastern Yorkshire. The time of the Ice Ages is described as Pleistocene, which extends throughout the Quaternary except for the post-glacial Holocene. The part of the Devensian that had the glaciers at their greatest extent is known as the Last Glacial Maximum. (MIS = Marine Isotope Stage.)

which ice remained on parts of Britain until 11,700 years ago. Though neither the longest nor the coldest glacial event, the Devensian glaciation had the greatest influence on the modern landscapes, simply because it was the most recent, so its landforms are relatively fresh, and are generally the most conspicuous.

The greatest Ice Age, when ice caps and glaciers reached their greatest extent, was the Anglian, dated at 480,000 to 425,000 years ago, and denoted as MIS 12. Anglian ice completely covered Yorkshire, and reached as far south as the outskirts of London. There are 104 stages in the complete isotope record, reaching back to 2,600,000 years, starting with the first one that was cold enough to induce significant glacial expansion. That's 52 cold stages, and probably twenty of those induced some degree of glaciation in Britain. But landforms of the first 46 were obliterated by the Anglian glaciation, when ice covered the Moors and the Wolds and all adjacent areas to depths of hundreds of metres. Moving generally southwards, typically by less than a metre per day, Anglian ice moved faster and excavated a little deeper along pre-existing valleys, so it did impact on the landscapes, but few individual features survived the long period of subsequent erosion.

Between Anglian and Devensian there were climatic oscillations, but there is considerable debate over how much ice ever spread across England during that interval between them. An intermediate cold stage is widely referred to as the Wolstonian, which appears to be aligned with MIS 8, or just possibly with MIS 6. Near the western end of the Vale of Pickering, the villages of Great and Little Barugh and Kirby Misperton (with its adjacent Flamingo Land theme park) each stand on small patches of higher ground. Originally they attracted their settlements because they were islands of dry ground within an extensive wetland. Those islands were, and still are, low mounds of glacial till rising above the alluvium and lake sediments that comprise the floor of most of the Vale. But their till is distinctive because it appears not to contain pebbles or rocks from any distant sources, as is typical of Anglian till that spread from Scandinavia and Scotland. So it seems possible that the till was locally sourced, and was carried there by Wolstonian ice flowing from a small icefield that developed on the Tabular Hills.

Devensian glacial till exposed on the foreshore at Mappleton, on Holderness, with blocks of white chalk conspicuous in a matrix that is dominantly clay carried from Jurassic outcrops.

By analogy there could have been ice on the Moors and Cleveland Hills, when any wider ice sheet from the north failed to surmount the high ground in eastern Yorkshire. But the evidence is not conclusive, and there are still no landforms in the Moors and Wolds that can be recognised for certain as originating between the two main glaciations of Yorkshire.

Devensian ice around the hills

During the most recent Quaternary glaciation, Devensian ice extended from Scandinavia across to Scotland. Its ice flowed slowly towards the south, but divided around the upstanding block created by the Cleveland Hills and North York Moors. The main ice flow was east of the hills, along the ground now occupied by the North Sea. However, sea levels were down by more than 100 metres due to so much water being held in the huge ice sheets of Scandinavia and Canada; in place of the North Sea, Doggerland was a bleak and cold lowland stretching across to Denmark. A divergent glacier flowed down the Vale of York, where it picked up tributary flows from the northern Pennines. North Sea ice also skirted the eastern flank of the chalk Wolds, and eventually wasted away on East Anglia, whereas the Vale of York Glacier only reached south of the Humber during a short-lived advance.

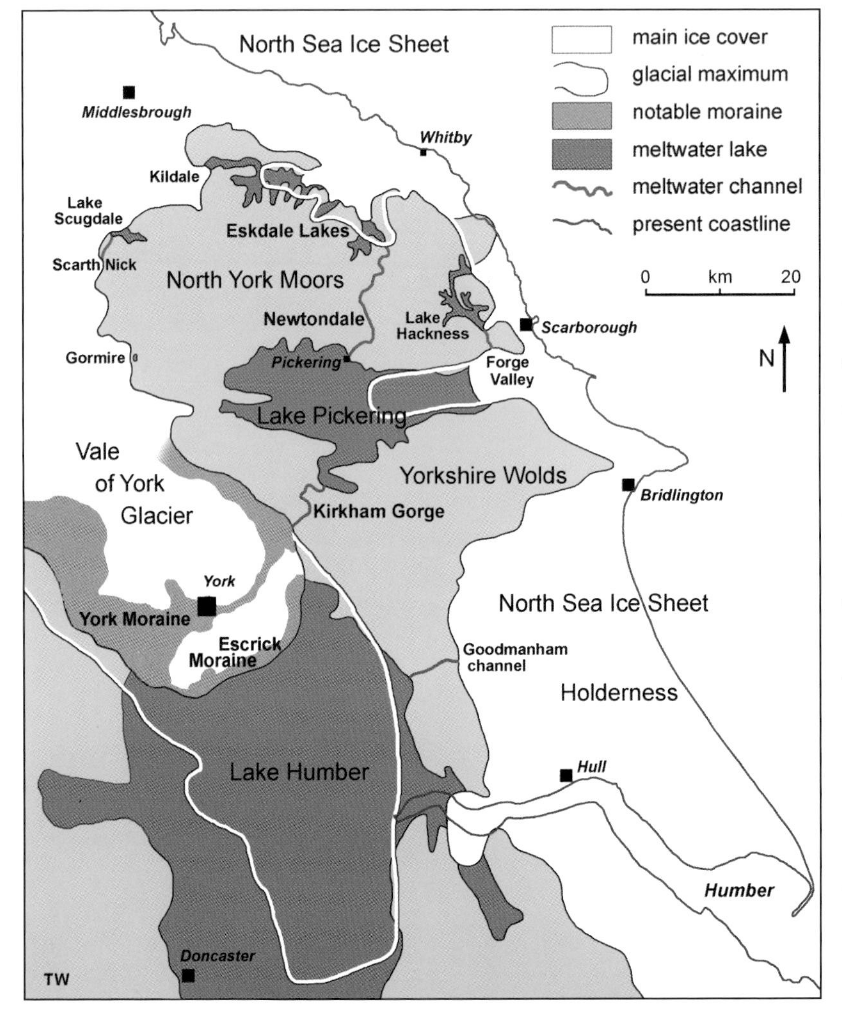

Eastern Yorkshire under Devensian glaciation. This map can only be a generalisation because it shows various features that changed considerably over a period of more than 15,000 years, from initial advance to final waning of the North Sea Ice Sheet. These are the major events that influenced today's landforms, but are not all contemporaneous. The ice cover is shown as it was after a modest decline from its maximum extent that is also shown.

Small glacial erratics, eroded by the sea from glacial till and recovered from Holderness beaches; each is about 60mm long. On the left: rhomb porphyry lava from the Oslo basin, carried to Yorkshire by Scandinavian ice. On the right: crinoidal limestone from the Pennines, carried to the coast by Scottish ice flowing through the Stainmore Gap.

Devensian ice never completely covered the Moors or the Wolds, but its influence reached beyond its margins, because it blocked the rivers flowing from the hills and thereby created a splendid series of ice-dammed marginal lakes.

North Sea Ice Sheet

Though barely recognisable within the farmed landscapes, there are patches of glacial till all along the coastal flanks of the Moors, where the ice spread up the lower slopes but never reached the high ground, though tongues of ice spewed into each of the valleys. Summer meltwater off the ice drained towards the exposed Moors, only to encounter drainage coming from the hills, where all the outlet valleys were then blocked by ice. The combined waters therefore accumulated between barriers of rock on one side and ice on the other, in what are aptly referred to as ice-dammed marginal lakes. Eskdale, the Upper Derwent and the Vale of Pickering held the largest lakes, but probably every other Moors valley held a lake for some length of time, until its water found a way out through the crevassed ice, or until the ice melted away when the world's climate warmed up. The lakes are now long gone, but the channels carved by their temporary overflows remain as deservedly famous features of the North York Moors (*see* pages 76, 78 and 84).

Becoming thinner, weaker and lower as it crept southwards, the North Sea Ice Sheet was deflected by the Yorkshire Wolds escarpment. It reached to elevations of nearly 150 metres across the chalk ridge that now reaches out to Flamborough Head, and then spread westwards on the lower ground to the south.

Holderness is composed almost entirely of glacial till, though there is hardly any exposure of it other than along the low coastal cliffs. A

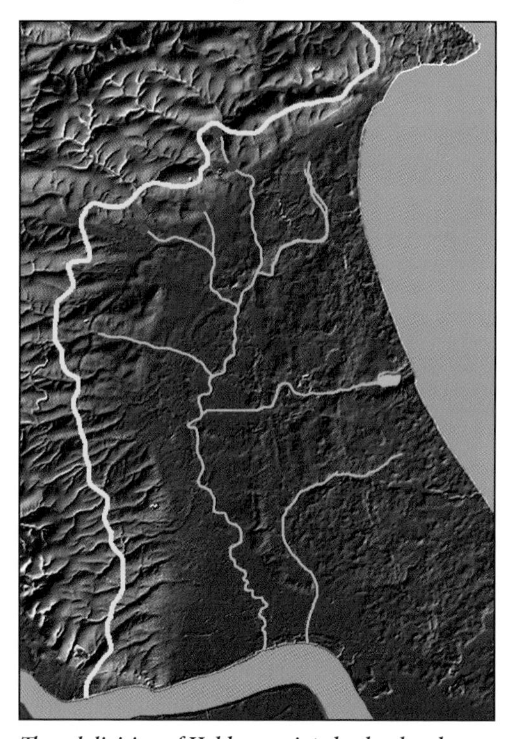

The subdivision of Holderness into lowland and flatland can be appreciated on an enhanced digital terrain model. The smaller area of flatland towards the west is on the alluvium of the River Hull, where the main river channels are shown in pale blue, with their largest springheads in dark blue. The larger eastern lowland shows by its more uneven terrain of glacial till. The yellow line denotes the edge of the till that was deposited where the ice sheet lapped on to the lower slopes of the chalk Wolds. The open sea and the Humber Estuary appear as dull blue.

Typical lowland of Holderness with farmland and lazy rivers, near Frodingham, southeast of Driffield.

The low ridge formed by the Kelsey Hill esker is now topped by a wind turbine that is sited on the highest ground amid the flatlands of southern Holderness.

sequence of three tills has been recognised. Youngest is the Withernsea Till, exposed only in the southern cliffs around Withernsea. The Skipsea Till is the most extensive, forming most of the cliffs, and a Basement Till is only exposed on the foreshore at Dimlington. The three tills look the same: all are clay rich and contain numerous erratics, large and small. These include blocks of chalk, of fossilferous mudstones from around Whitby, of strong rocks from northern England, and just a scatter of recognisable igneous rocks from Norway. Separating the tills are lenses of sand and gravel containing fragments of Arctic plant species. The three tills are interpreted as the result of successive advances of a surging ice sheet, in which flow periodically accelerated with no immediate link to climatic fluctuations.

The topography of Holderness is not entirely flat. Its eastern and larger part is uneven ground typical of glacial moraine, with local relief of up to twenty metres. Foston on the Wolds is a little village, east of Great Driffield, which stands on a ridge of glacial till only a few metres high, but enough to keep the houses above the surrounding wetland. The village name derives from a wold being an upland with no trees, but the site is not a wold in the

A headland cliff-top just south of Flamborough Head; North Sea Ice moved in from the right over frozen bedrock chalk, which had its upper layers dragged into a small fold; this can be seen to be more overturned upwards until its structure is lost in the increasingly frost-shattered ground.

sense of the nearby chalk Wolds. Different again is Kelsey Hill, much further south, and east of Hedon: this is an esker, which is a strip of sediments deposited by a sub-glacial river; it is therefore composed of gravel, though nearly all of this has now been quarried away.

Along the length of eastern Holderness there is a crude texture of low, north–south ridges that appear to be formed of till crumpled up by glacial surges pushing ice outwards on to the lowland. These ridges may have been formed by ice bulldozing the till into banks, but can also be due to ground deformation where glacier ice is almost welded to the underlying frozen ground, which is then distorted as it is dragged along by the inevitable flow of the ice. The latter process is known as glacial drag folding, which can also distort upper layers of weak bedrock; this is clearly exposed in the sea cliff south of Flamborough Head, where the chalk has been folded by ice moving over it from the north.

A major slice of western Holderness has noticeably flatter ground that has been levelled out by alluvium deposited on all but the highest parts of the glacial till. Draining south along its axis, the River Hull meets the sea at the city of Hull, which is correctly named Kingston upon Hull. Main branches of the river are fed by large springs where groundwater in the chalk breaks up through the thin edge of the glacial till. The river also captures most of the drainage from the eastern half of Holderness because the highest ridges within the deformed till lie almost along the coast. Hornsea Mere drains inland to the west, though the coast lies only a kilometre to the east. The Mere is the last of the many small lakes that had ponded on the irregular surface of the till; all the others were drained long ago to increase the area of farmland.

Extensive wetlands along the middle reaches of the River Hull are known as carrs. Most of these have been drained simply by digging straight channels to accelerate run-off, with their gradients steeper than those of the longer, meandering, natural channels. However, part of Leven Carr has recently been restored to wetland in order to improve the ecology and wildlife of the region.

Hornsea Mere, the remnant lake among drained wetlands.

The Rudston Monolith

There is still some doubt over one other probable relic of the North Sea Ice Sheet. The tallest standing stone in Britain is in the churchyard of Rudston village, near the eastern end of the Wolds. A single metre-thick bed of sandstone stands nearly eight metres tall, and there must be a few more metres of it underground; less than two metres wide, it makes an impressive needle. The monolith is an unusually large slab of Moor Grit, from the upper part of the Ravenscar sequence in the North York Moors.

Its placement is commonly ascribed to the Bronze Age, around 2500 years ago. However, it is hardly realistic to think of the local people dragging it at least 30 km from the nearest outcrop of Moor Grit in the coastal hills around Cloughton, north of Scarborough. On the contrary, it is very plausible that such a block could have been plucked from the outcrop by Devensian North Sea ice, and left as an erratic somewhere close to Rudston when the ice wasted away across the end of the Wolds.

An alternative source is the main southern outcrop of Moor Grit across Spaunton Moor, northwest of Pickering. This is also reasonable, as noticeably large slabs of the grit do occur there, one of which has become the Millennium Stone now standing high on Danby Moor (*see* page 20); however, that was taken to site by tractor and trailer, not on Bronze Age rollers. Furthermore, if the Rudston monolith is a glacial erratic from Spaunton Moor, it would have to date from the Anglian glaciation, as Devensian ice never flowed that route, and it is also questionable that an ice sheet would have carried a large erratic up the scarp face of the chalk hills to end up anywhere near Rudston.

As far south as the Yorkshire Wolds, the North Sea Ice Sheet was always rather thin, and wasted away long before any serious decline of the thick icecap on Scandinavia. So there was no Yorkshire coast while Doggerland was clear of ice, and sea level was still 100 metres below that of today. In place of the southern North Sea there was a cold, barren, recently de-glaciated lowland extending far to the east of Holderness – and that could well have been a significant source of loess carried on to the chalk Wolds whenever strong, dry winds blew from the east. There was probably dry land reaching many kilometres east of the present coastline until sea level rose rapidly by about three metres around 8200BP (or 6200BC) when the enormous ice-dammed Lake Agassiz drained from the Canadian Prairies into Hudson Bay.

It was probably that event that caused the final evacuation of primitive communities from the last remnants of Doggerland. A mass exodus was timely, because it was probably just a few years before all coastal areas were devastated by a tsunami originating from a huge underwater landslide at Storegga off the coast of Norway. The North Sea has its own story to tell, though there is very little that survives from the people who once lived on Doggerland. That terrain is now lost beneath the waves – yet still the North Sea continues to expand by eating its way into Holderness, as it slowly removes a lowland of glacial till that was only created little more than 10,000 years ago.

Glacier in the Vale of York

Through most of the few million years that the Moors and Wolds have been their own landscape units, there has been open sea along their eastern margins. The Vale of York has remained above sea level, but it was occupied by ice during the Devensian glaciation. The Pennines, too, were then free of ice, but a glacier 30 km wide crept southwards down the Vale as an offshoot from the North Sea Ice Sheet.

For most of that time the glacier ended where it was, calving chunks of ice to form icebergs into Lake Humber. That bygone lake existed when meltwater from the glacier, along with drainage from all the adjacent hills, could not escape out through the Humber Gap because it was blocked by the North Sea ice. The waters ponded up behind the ice barrier until Lake Humber reached an altitude of about 42 metres and extended far to the south, where it found an outlet though the Lincoln Gap. The lake held its maximum size for a few thousand years before its level started to fall when an exit through the Humber Gap became available as the North Sea Ice Sheet slowly declined. During that phase, the Vale of York Glacier appears to have advanced briefly as far south as Doncaster. However, the effects of those events are now largely obscured by the extensive alluvium and peat bogs that occupy the lowlands around Goole.

Within the Vale of York, the two well-known landforms are the moraines of Escrick and York that form low, arcuate ridges across the flatlands. These are both classic end moraines that were left during the retreat of the glacier.

The centre of York stands on its moraine that originally provided a raised patch of dry ground beside the River Ouse amid wetlands extending north and south. The moraine is barely recognisable within the urban area, except when frequent flooding inundates the riverside buildings that stand on low ground where the river had previously eroded into the vital ridge of glacial till. However, it is

The great flatland of the Vale of York, seen from the Wolds north of Pocklington.

The road from Howden to York curves to the left in front of the grassy ridge formed by the York Moraine.

easily seen where the road to Hull (A1079) lies along its crest and crosses smoothly over the newer bypass (A64) that lies on the level of the alluvium some ten metres below. The actual height of the moraine is more than ten metres, as that is just how much it projects above the alluvium surface. Along with lacustrine sediments, the alluvium varies greatly in thickness, but can reach to many metres where it has filled valleys in the pre-glacial landscape to create the wide, flat floor of the current Vale.

Glacial Retreat and Moraines

Terminology can be confusing here. A glacier does not retreat, in that it does flow back from whence it came. Instead, as climates ameliorate, the rate of onward ice flow becomes less than the rate of melting at its terminus; so the terminal ice face retreats even while ice still flows down its valley, or in some cases reaches a state of stagnation.

At the end of the glacier, rock debris carried by the ice is dumped, to form a moraine as the ice melts away. A moraine is a landform, either a sheet or a ridge, made of till. Moraines can be formed at the ends of stable glaciers, but many are enlarged due to climatic oscillations when a glacier retreats a little and then advances again, bulldozing the till into a taller moraine. Any significant glacial advance wipes out older moraines, and a significant landform only survives beyond a retreating glacier. So the ridges of debris across valleys such as the Vale of York have long been known as end or terminal moraines, though are now correctly referred to as retreat or recessional moraines.

A lateral moraine forms along the margin of a valley glacier, mainly by collecting debris that falls from weathering rock slopes that overlook the moving ice.

A low ridge of glacial debris pushed up at the front of a glacier to form a retreat, or terminal, moraine; this one is at Exit Glacier in Alaska.

The River Ouse meanders across the broad flatlands around Goole, on what was the alluviated floor of Lake Humber that extended south from the front of the Vale of York Glacier in Devensian times.

The peat bogs are the youngest features of the Vale, generated in the extensive wetlands. Much of this ground has been drained to increase agricultural values, and some of the peat has been extracted for horticultural use. Current appreciation of environmental values has led to restoration of some areas of wetland, along with closure of the peat extraction sites.

The Escrick Moraine is rather overshadowed by its York neighbour, as it forms a low ridge barely distinguishable within the landscape, though it does keep the villages of Bolton Percy, Escrick, Wheldrake and Newton upon Derwent clear and dry above potential flooding on the alluvial flats. Both the York and the Escrick moraines were formed during glacial retreat along the Vale around 16,000 years ago, with the Escrick pre-dating the York probably by only a matter of decades.

A lateral moraine along the Moors bank of the Vale of York cannot be described as a major feature of the landscape, though it does rise enough to form higher ground between the villages of Easingwold and Crayke. Further north it blends into the sheet of till that was dumped by the glacier wasting away on the flatlands between the Cleveland Hills and the Pennines.

When it was at its maximum depth, the Vale of York Glacier extended a lobe of ice into the Coxwold Gap between the Hambleton Hills and the Howardian Hills. As it melted back, this distributary glacier deposited a bank of till to create the low Ampleforth Moraine. For a limited time this probably formed a short western shoreline of Lake Pickering (*see* map on page 66), and its remains survive to form the watershed within the rift valley. But ice encroached no further on to the Moors and Hills, and the remaining story from the Ice Age is one of meltwater that was both flowing and ponded.

Meltwater rivers and lakes

During the Devensian glaciation, the Cleveland Hills and North York Moors stood high enough to avoid being overrun by ice sweeping down from the north. Ice flowed south along both sides, while the hills and moors stood clear as a wild, desolate and very cold island of tundra. With the ice effectively forming new ranges of hills, albeit short-lived, the drainage systems of the Moors were temporarily forced into new alignments. After the ice had melted away, around 12,000 years ago, the rivers returned to their original courses, though there were a few exceptions. But the erosional imprint of those Ice Age meltwater rivers remains as a conspicuous component of the modern landscape.

Glacial lakes of Eskdale

The River Esk gathers run-off from much of the northern part of the Moors, as it has done throughout a long history of landscape evolution during the Quaternary. But its situation changed when Devensian ice almost surrounded the hills and blocked the valley's outlet to the sea at Whitby. As the North Sea Ice Sheet thickened, it fed a small glacier into Eskdale, heading up the valley, even reaching west of Lealholm for a time. Snowmelt and rainfall continued to pour from the hills, and ponded up within the ice-dammed dale. Eventually it overflowed at the lowest point on its margin, and a powerful river headed southwards, and scoured out Newtondale. This

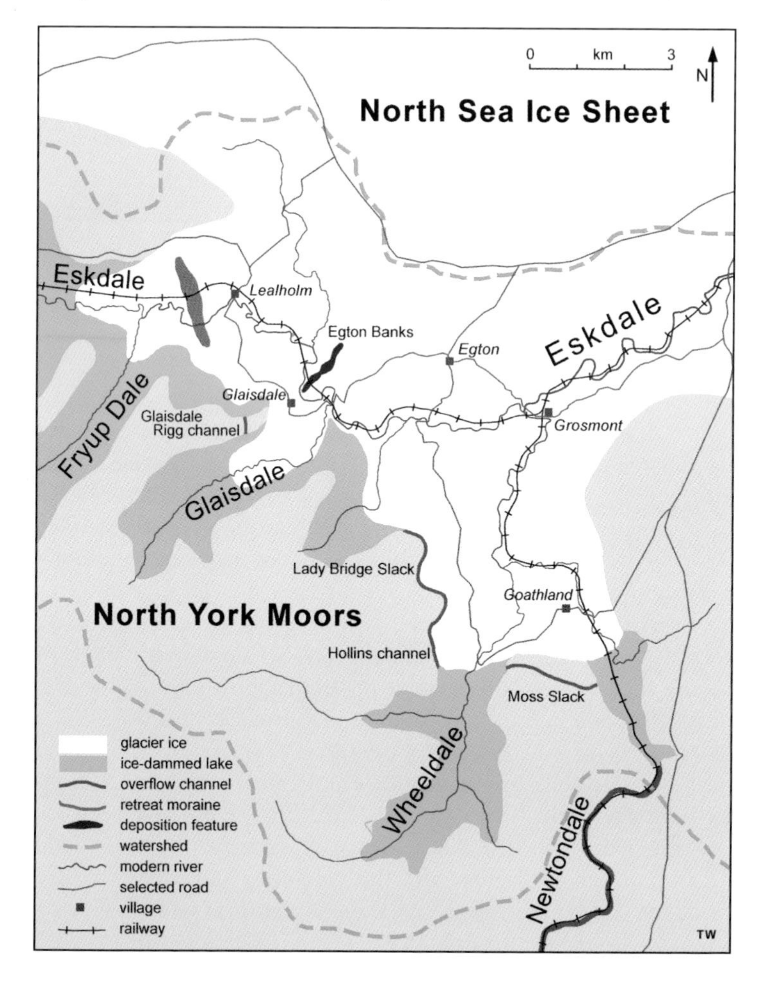

Some prominent landforms were formed in the central section of Eskdale during an early stage in the Devensian evolution of its glaciers and ice-dammed lakes. In upper Eskdale, west of Lealholm, the glacier had previously extended further to the west, and its retreat subsequently paused at the position of the moraine that is shown in position, though then not yet in existence. Moss Slack and the other overflow channels were probably active throughout that period of shrinkage in the glacier west of Lealholm.

has carried no significant stream since the ice melted away, when the River Esk resumed its course out to Whitby.

In 1891, Professor Percy Kendall, from the embryo Leeds University, recognised Newtondale for the overflow channel that it had been, and pursued his studies upstream to identify the extent of the lakes that had once existed in Eskdale. Between times he visited the Marjelen See in Switzerland, a beautiful lake ponded against the great Aletsch Glacier, where he made comparisons, and then in 1902 published his classic paper on the 'Glacier Lakes of the Cleveland Hills'. This was a landmark first description of the glacial diversion of rivers, where meltwater created channels, many of which were overflows from ice-dammed lakes, and all are complete anomalies within the modern landscapes. Eskdale was his subject for research, and his principles were then applied elsewhere within Britain's glaciated landscapes.

Lake Eskdale was not a singular feature. Things started happening, probably soon after 26,000 years ago, when the Scandinavian–British ice sheet expanded southwards to displace the North Sea and then block off the outlet of Eskdale. A lake formed, with its waters backing up until they found an outlet over a col at the head of the Eller Beck valley, south of Goathland, and into Newtondale. The greatest extent of the lake was achieved when the col was first crossed at an altitude of a little under 230 metres. Subsequently, overflow waters cut

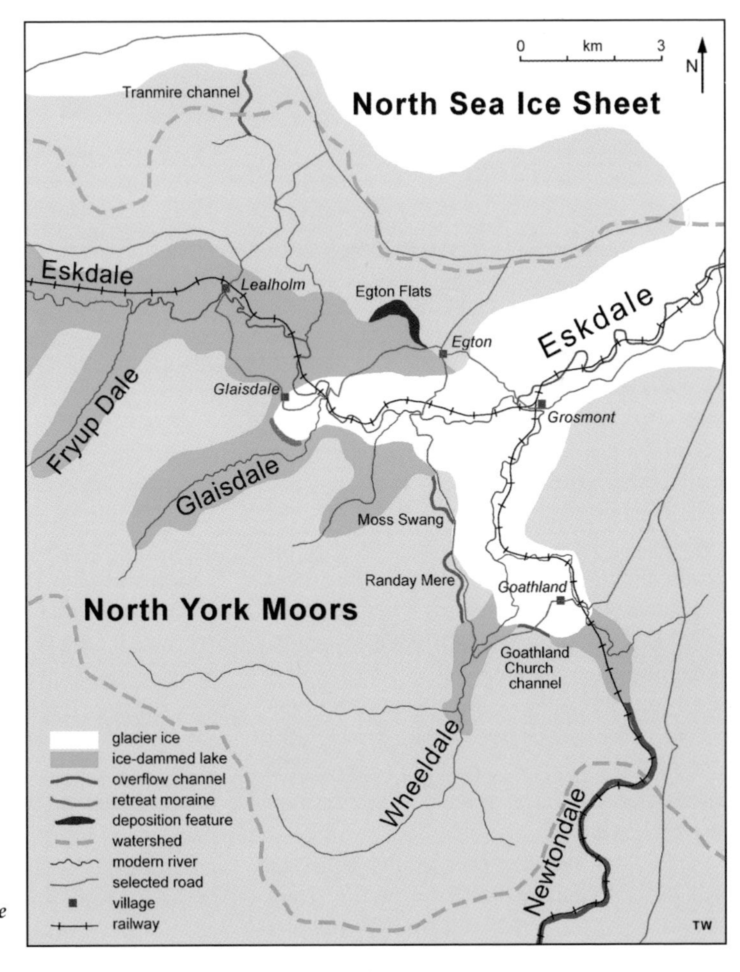

Some of the more conspicuous features that were formed in the central section of Eskdale at a later stage in the Devensian evolution of its glaciers and ice-dammed lakes. Moss Swang and the other overflow channels are at lower elevations than the older channels shown on the map on the opposite page. In its relationship to the other marked features, the precise timing of the retreat moraine in Glaisdale remains uncertain.

Newtondale

Now accessible in comfort on the steam-hauled trains of the North Yorkshire Moors Railway, Newtondale has to be the most magnificent meltwater channel in Britain. It was formed almost entirely during the few thousand years when the North Sea Ice Sheet blocked the outlet of Eskdale towards Whitby, thereby impounding a temporary Lake Eskdale. That lake overflowed at the lowest point on its rim, so meltwater then poured down the line of Newtondale. Flow rates varied enormously: high in the relatively warm summer, low when both lake and landscape were frozen in winter, and highest of all in spring when snow melted away from the unglaciated hills. Flows also varied every day, from early morning lows to late afternoon highs, due to the power of solar radiation in daytime.

The end result was a 15-km-long channel, mostly around 300 metres wide and in places 100 metres wide. It winds southwards across the Moors, almost certainly following an earlier stream course, which already had a narrow valley through the Tabular Hills, and this, too, was enlarged by the meltwater invasion. When the North Sea ice retreated and Lake Eskdale drained away, Newtondale was abandoned. It is now drained by the very modest Pickering Beck, which is an underfit stream with a much larger ancestor that created its valley.

down into the weak and weathered sandstones and mudstones of the Scalby Formation that formed the watershed. The level of Lake Eskdale progressively fell from its short-lived peak to that of a more stable outlet, which was eventually lowered to the present level of about 165 metres, where the Fen Bogs now stand on the col at the summit along the railway. At its maximum, the lake extended across the low saddle into Kildale, where the westward outlet was also blocked by ice. Lakes evolve with time; even the Marjelen See disappeared in the 1980s when its water drained away through the Aletsch Glacier.

As climates became ever cooler into the Last Glacial Maximum (referring to the extent of the ice), the North Sea Ice Sheet became thicker, so that a glacier peeled off into Eskdale, maintaining a downward gradient while expanding upstream along Eskdale. This was joined by ice flowing over the watershed on the northern side of Eskdale, crossing the high ground north of Egton. The combined result was a glacier that extended well up Eskdale, greatly reducing the size of Lake Eskdale. With the Eskdale Glacier at its maximum, drainage from the ice-free Moors, along with summer meltwater from the ice, accumulated in a chain of lakes occupying each of the tributary valleys, including Westerdale, the Fryup Dales, Glaisdale and Wheeldale. Their overflow waters worked their way eastwards along the ice margin until they found the outlet over the col into Newtondale.

The eastern arm of Westerdale, seen from Castleton Rigg, is one of the Eskdale tributary valleys that briefly had Devensian ice moving up it, before glacial retreat left it to be occupied by an arm of Lake Eskdale.

Water levels in these temporary lakes of the past were rarely stable for periods long enough to create wave-cut notches or terraces of beach sediment that can be recognised along today's hillsides. Instead, Professor Kendall identified the lakes by tracing the channels that were carved by their overflow waters before being abandoned to remain as anomalies in the modern landscape. These overflow routes can take any of three forms.

Kendall described the direct overflow channel as one that has bedrock slopes along each flank. Most of these are short, across shoulders and interfluves, though Newtondale is the grand exception, reaching 15 km across the Moors and Tabular Hills. Marginal overflow channels were Kendall's second type, formed by water flowing between one flank of rock hillside and an opposite flank formed of ice at the glacier margin; after ice retreat, these are left as asymmetrical channels with all or much of one side missing. Later studies of glaciers have recognised a third type of lake outlet that was inside or beneath the ice, and many channels, in Eskdale and elsewhere, have now been re-interpreted as sub-glacial features.

During the wane of the Devensian Ice Age, it took some thousands of years for the Eskdale Glacier to decline and disappear, with climatic oscillations alternating short phases of stability and retreat. Ice surface altitudes varied by many metres across the slope of the active glacier, and the grand retreat left behind isolated patches of stagnant ice that slowly melted away. So it is indeed difficult to identify all the stages in a complex deglaciation of Eskdale.

Moss Slack, a classic overflow channel cut across a moorland ridge by meltwater from Wheeldale draining eastwards into Newtondale, when both dales contained lakes dammed by the Devensian Eskdale Glacier.

Eskdale's overflow channels

The maps on pages 74 and 75 are mere snapshots of two stages within the evolution of the glacier and lakes in the central part of Eskdale. Each map marks just some of the more conspicuous landforms that might be roughly contemporaneous.

That of the earlier stage shows features that developed after an Eskdale glacier had already declined far from its maximum, when it had briefly reached as far west as Danby. South of Goathland, Moss Slack is perhaps the finest of the high-level overflow channels, easily reached from roads near both ends, and with a small tarn on its 30-metre-wide flat floor towards its western end. 'Slack' and 'swang' are both terms derived from Old English for low wetland channels. Moss Slack drained out of Lake Wheeldale, which had its own input along Lady Bridge Slack and around one side of Hollins Hill. This combined drainage route was essentially an ice-margin channel, with parts of it cut into bedrock beside the ice, and

Moss Swang, a low-level overflow channel that once carried water from Lake Eskdale towards Wheeldale and the Newtondale outlet, when a shrunken lobe of North Sea ice still reached as far as Goathland.

The magnificent Tranmire overflow channel, formed when meltwater from the North Sea Ice Sheet was ponded on its margin against the northern rim of the North York Moors until it overflowed across the watershed and down into Eskdale.

parts that were in or under the ice sheet. So it is not so well defined, and has gaps along its line, either where flow was through tunnels within the ice or where a shallow channel has been removed by subsequent erosion.

A stream cave inside a glacier is very capable of moving sand and gravel along its bed. Sediments may be left inside a glacier cave when the stream ceases to flow, and are then left behind when its host glacier becomes stagnant and melts away. That strip of sediment becomes an esker, an irregular ridge of gravelly debris that is yet another anomaly within the modern terrain. The twenty-metre-tall, grassy ridge of Egton Banks is such an esker, formed beneath the Eskdale Glacier while this section was wasting away. The more gently graded ridge west of Lealholm appears to be a retreat moraine, formed a little later when the toe of the glacier had melted back from its mapped position and briefly stabilised to dump the mound of unsorted till.

The Egton Banks esker, formed of glacio-fluvial sands deposited inside a glacier cave when Eskdake was occupied by Devensian ice.

The second map shows Eskdale in a later stage with a smaller glacier impounding lakes at lower levels but still draining out to Newtondale, which had by then been entrenched to almost its present depth. Moss Swang is another classic overflow channel, though its equally deep, downstream continuation is shrouded in forest around Randay Mere. By this stage, the North Sea Ice Sheet had waned to a lower level so that it only surmounted Eskdale's northern ridgeline far to the east of its earlier extent. Meltwater still accumulated along its margin, and at one place escaped over the watershed to form the Tranmire channel. This is the third of Eskdale's classic overflow channels, and is especially fine because it crosses a major watershed.

Egton Flats is the name for a broad, dissected terrace that is very noticeable with its cover of farmland pasture. Its position suggests that it could be a shoreline feature, but it is equally likely to be a kame terrace formed when the glacier extended in front of it. A kame terrace is formed of stream sediments deposited along the margin of a glacier, then left as a bench along a hillside when the glacier has melted away.

The high-level terrace known as Egton Flats is probably a kame terrace that was formed when Devensian ice occupied much of Eskdale and meltwater drainage left these sediments along the glacier margin.

Lake Pickering

Surrounded by hills, with meltwater pouring in from all sides, and with a wall of ice blocking its outlet to the North Sea, the Vale of Pickering had to become a lake through much of the Devensian Ice Age. It could have filled to a maximum level of around 70 metres, early in the Devensian when ice also blocked both the Coxwold Gap and all lower cols across the Howardian Hills around Malton, though there is still some doubt over details of that initial stage. With the glaciation at its maximum, a tongue of North Sea ice may have reached along the Vale nearly as far as Pickering, calving icebergs into the lake that was steadily gaining in level. Till remains buried beneath the lake sediments.

However, lakes are ephemeral features in any landscape. Their demise is only a matter of time when they are progressively drained by river erosion that lowers their outlets while at the same time they are filled by sediment carried in from any number of inlets. Though Lake Pickering may have extended towards the east as the Vale's lobe of North Sea ice retreated, the continuing story of the lake was one of steady decline, with its size reducing as its level fell.

Around 23,000 years ago, the lake level was at about 45 metres (above today's sea level), when its eastern end was against the ice and its debris, the latter still recognisable as the Wykeham Moraine (roughly as shown in the map on page 83). A northern part of this retreat moraine forms the low ridge on which Wykeham Abbey has long been situated, whereas the southern part of the moraine was lower and now lies buried by later lake sediments. The bank of glacial debris extends to the northeast, where the villages of Hutton Buscal and East Ayton stand on its slightly raised ground. This is often described as a kame terrace, formed of sediments deposited by meltwater flowing along the edge of the glacier, but parts of it are more like a lateral moraine, and it is quite typical to have a mixture of glacial and fluvial debris at a glacier margin.

By this time, the outlet of Lake Pickering was established on the line of the Kirkham Gorge, which had been initiated where the lake overflow found a route through the Howardian Hills and out on to the end of the Vale of York Glacier. Fluvial erosion continued to deepen this channel, and thereby drain the lake to progressively lower levels. In the easily eroded bedrock of the Howardian Hills the channel has widened out so that it hardly warrants the label

The eastern part of the Vale of Pickering with the Tabular Hills on the skyline.

of a gorge, though it is a conspicuous element of the regional landscape. It still carries the River Derwent, on the banks of which stand the skeletal remains of Kirkham Abbey.

Over the next few thousand years, lake levels continued to decline. Hillside terraces that combine wave-cut notches with beach accumulation are the classic way of recognising ancient lake levels, but they are very difficult to identify in the Vale. This is not Glen Roy, that famous valley in western Scotland with the 'Parallel Roads' clearly seen along its slopes, even though both Vale and Glen contained ice-dammed marginal lakes during the Devensian. Glen Roy had a succession of stable lakes separated by major changes to its ice dam, whereas the lake in the Vale of Pickering fell slowly and steadily as its outlet river deepened the Kirkham Gorge.

The River Derwent flows through the rather open valley known as the Kirkham Gorge, which was formed when the ice-dammed Lake Pickering overflowed into the Vale of York.

By about 18,000 years ago, Lake Pickering reached westwards barely beyond its Kirkham outlet, whereas it had extended eastwards to the edge of the lateral moraine left by the shrinking North Sea Ice Sheet, and on which now stand the villages of Cayton, Lebberston and Gristhorpe (*see* map on page 83). Subsequently the lake level fell even further and its expanse of open water was replaced by a huge wetland of marshes and small lakes, where Lake Flixton was probably the largest remnant to survive for longer at the eastern end of the Vale. After the ice had all gone, that low barrier of moraine stretching from Scarborough to Filey and beyond was still enough to ensure

Thick glacial till that has been scored by rainwater gullies on Filey Brigg; this is a part of the lateral moraine left by the North Sea Ice Sheet.

that no post-glacial river could drain from the Vale of Pickering eastwards to the sea. The River Derwent became the new trunk stream, flowing out through the Kirkham Gorge.

The great inheritance of Lake Pickering was the spread of sediments that had accumulated within it, and which now form the floor of the Vale. Fine-grained, lacustrine silts and clays underlie all the flat areas, commonly beneath a veneer of later alluvium or peat. In contrast, the town of Pickering stands on a great sand delta that was built out into the lake by the meltwater pouring down Newtondale; it rises above the flatlands just the little needed to keep the town dry and clear of undrained marshes. Original settlements within the Vale were sited on islands amid the wetlands prior to any drainage schemes developed by subsequent inhabitants. These islands are the crests of low hills of either pre-Devensian glacial till or Kimmeridge Clay that projected above the fill of lacustrine sediments; Kirby Misperton stands on the largest.

One settlement stood on a fragment of barely perceptible sand delta that extended south from the meltwater exit from Mere Valley and rises only just clear of the wetland. This was Star Carr, which benefited from its site on the western shore of Lake Flixton before it was choked with vegetation. Star Carr ranks as the finest Mesolithic site in Britain, and among the most important in Europe, due to the unusually well preserved state of the numerous artefacts that have been excavated from beneath a metre or so of clay, peat and soil. It was first occupied around 9300BC. That was only a few centuries after climatic improvement heralded the final demise of glaciers in Britain, though North Sea ice had gone from the Yorkshire coast a few thousand years beforehand. Excavations at Star Carr started in the 1950s, and intermittent works ever since have quickly been restored to farmland, so there is no visitor access or facilities, despite its historical significance.

Carrs are the Yorkshire version of fens, wetlands that commonly stand on peat reaching to a few metres thick. The names of carrs appear on maps of the Vale of Pickering, mostly of the eastern half where Lake Pickering lasted longer, though nearly all have been drained over the centuries to create productive farmland. It is only recently that the natural values of wetlands have been more fully appreciated, with restoration of selected areas now underway. Parts of Flixton Carr and Cayton Carr are once again thriving wetland, giving some indication of how the Vale of Pickering looked in its natural state, after the glaciers had left their mark and before evolving further under the influence of mankind.

Amid the flat wetlands of the Vale of Pickering, a low hill of very old glacial till projects through the lacustrine sediments to provide a dry site for the village of Great Barugh.

Diversions of the Derwent

East of Newtondale, a meltwater channel with a different history is the Forge Valley, which is still occupied by the River Derwent. Both the Dalby Forest section of the Tabular Hills and an eastern portion of the Moors drain into the Upper Derwent, which originally flowed to the sea at Scalby, immediately north of Scarborough. Then, like Eskdale, this outlet was blocked by the North Sea ice. Water backed up within the valley to form Lake Hackness, with its level rising to the lowest col on the southern ridge of Calcareous Grit, at which point the overflow escaped southwards into the Vale of Pickering, thereby carving out the Forge Valley. Lake Hackness declined in size as the floor of its Forge Valley outlet was progressively lowered; there is little clear evidence as to how long the lake lasted at its various levels. The map below shows the lake at close to its maximum size, whereas a shrunken late stage is depicted in the map on page 66.

As climates ameliorated towards the end of the Devensian, the ice dam in the Derwent Valley disappeared, but left behind a low bank of glacial till, a small retreat moraine, across the valley. By that time the Forge Valley had already been cut to a lower level, so the River Derwent never reverted to its pre-glacial course. It still flows through the Forge Valley, beside a minor road, in a wooded ravine that is two kilometres long and 70 metres deep at its upper end.

Once clear of the Forge Valley, the Derwent heads first south and then west along the Vale of Pickering, before heading off on its remarkably circuitous route eventually to reach the sea through the Humber Estuary. This massive glacial diversion has left the River Derwent with nearly 100 km to travel from an altitude of only 30 metres at its exit from the Forge Valley. Frequent and extensive flooding along many parts of its lowland course have been an inevitable consequence, whenever peak flows could not drain away fast enough on the minimal gradient.

ABOVE ***The road across Folkton Carr stands on the soft sediments of glacial Lake Flixton. Repairs have been needed where its approach to the River Hertford crossing is slowly subsiding adjacent to the bridge that is stable on deep foundations.***

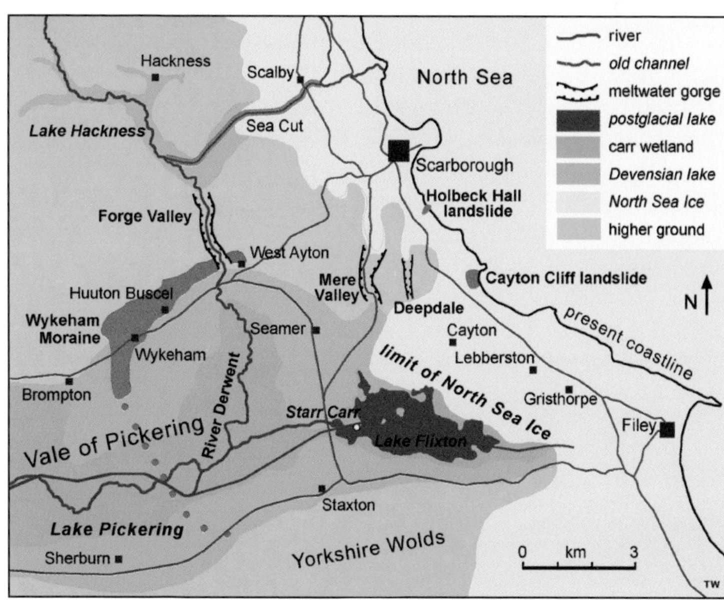

RIGHT ***Major landforms of the eastern end of the Vale of Pickering and along the upper River Derwent. Features labelled in italics no longer exist.***

The Sea Cut carrying floodwaters from the Upper Derwent out to sea north of Scarborough.

A major flood in 1799 prompted excavation of the Sea Cut, which was opened in 1804. This takes water from the Upper Derwent almost back on to its pre-glacial course, with an engineered channel 5 km long carrying floodwater from the Derwent to an outfall into Scalby Beck, which descends steeply to the coast. When floodwater is directed into the Sea Cut, a base flow is maintained on the Derwent, to continue through the Forge Valley. Across the Vale of Pickering the river once took a meandering course, but sections have been canalised in order to shorten the path, increase the gradient, accelerate the flow, and thereby reduce flooding.

East of Forge Valley, the Weaponness Valley was another breach in the Tabular Hills, which North Sea ice pushed into when at its maximum extent. A narrow glacier extended in from the north, and meltwater contributed to enlarging its southbound outlet, also known as the Mere Valley. There is now no river through the valley, but its floor holds The Mere, the remnant of a natural lake, and carries the railway and main road south out of Scarborough. East of Mere Valley, Deepdale is yet another, even smaller, channel that carried meltwater from the North Sea ice, probably for only a short time.

Western meltwater channels

When they were an island of rock almost surrounded by ice, the North York Moors had meltwater streams, ice-dammed lakes and overflow channels on all sides. In addition to the grand landforms on the eastern margins, there are smaller features along the western side that overlooks the Vale of York. In its northern corner of the Cleveland Hills, Kildale contained an ice-dammed lake, but for much of its lifetime this was a western arm of Lake Eskdale, so drained out through Newtondale.

Next to the south, a tongue of ice extended well into Scugdale for a short time, whereas a small lake probably existed for much longer. Its overflow carved out the Scarth Nick channel, which drained back to the ice margin at Osmotherley, and now contains the Cod Beck Reservoir. This is a broad, open channel that probably carried meltwater for a substantial length of time. Rather different is Holy Well Gill, a parallel channel two kilometres to the east, which is a narrow, descending gulley some ten metres deep with steep sides. Its upper end is around 70 metres higher than Scarth Nick, so it is likely that it carried overflow from a high-level Lake Scugdale for just a short time, before water found a lower exit through Scarth Nick.

South of Osmotherley, fragments of bygone meltwater channels can be identified within the modern landscape. A glacier presents an active and ever-changing environment, and marginal channels can be excavated and abandoned at a pace, especially where streams can loop through tunnels within the ice, of which no trace may survive. One channel fragment lies below Whitestone Cliff at Sutton Bank, and contains the little lake of Gormire; the ends of the lake are ponded against banks of landslide

Scarth Nick breaches a watershed where it was carved out by meltwater draining through lakes along the margins of the Devensian ice in the Vale of York.

debris that slumped from the mixed sandstone and mudstones beneath the high crag, long before the rockfall of 1755. Continuation south of this marginal channel can be recognised as the deep saddle between Roulston Scar and the outlier of Hood Hill, then along a smaller depression from Oldstead to Byland Abbey, heading towards the Coxwold Gap. Outfall was into a contemporary Lake Pickering.

Beyond the end of its glacier, the Vale of York had another channel draining into it, carved by meltwater from the North Sea Ice Sheet that was resting up against the far side of the Yorkshire Wolds. Goodmanham Dale cuts right across the crest of the chalk escarpment, reaching for about six kilometres from an open head on the dip-slope to its exit through the scarp face at Market Weighton. It could have been fed by streams on or inside the ice, which just found a convenient way out across a col in the low chalk hills. Alternatively it could have been overflow from a marginal lake that sat between the ice and the hills; there is, however, little indication of how extensive such a lake might have been, and it could have existed for only a short time. The dale is now dry, as it would be anyway on the unfrozen chalk, but it has proved to be a convenient low-level corridor for a minor road, for a railway now long gone, and for a pipeline carrying North Sea gas from the Easington terminal.

A link from Ice Age meltwater to North Sea gas may seem obscure, but it is there within the landscape of the Yorkshire Wolds.

Holy Well Gill was an overflow channel only active for a short time, as the outlet from a high-level lake dammed by ice and backed up into Scugdale.

Gormire, the sheltered lake within the high-level, Devensian meltwater channel along the eastern side of the Vale of York.

Goodmanham Dale, the abandoned channel that was cut by meltwater from the North Sea Ice Sheet draining across the Wolds and into the Vale of York.

CHAPTER 8

An Evolving Coastline

Between the Tees estuary and Flamborough Head, the coast of eastern Yorkshire provides a geological cross-section of the Moors and the Wolds, albeit one that is rather disjointed. Jurassic sandstones and mudstones of the Moors are exposed along the cliffs from Redcar to Filey Brigg, and the Wolds chalk is conspicuous from Bempton to Flamborough, though most of the intervening succession lies hidden beneath glacial till around Filey. Further south, the Holderness coast reveals only glacial till, but there the geological interest is in the erosion of a retreating coast – or perhaps that should be perceived as an advancing sea.

The Wrack Hills undercliff west of Runswick Bay, where an ironstone mine was destroyed in 1858 by renewed movement of the landslip terraces. Prior to that, the area had gained is name when seaweed, locally known as wrack, was taken from the foreshore and laid out to dry on the undercliff terraces before being spread as fertilizer on the fields above.

Yorkshire's Jurassic Coast

A long and wide strip of Jurassic rocks crosses England from Yorkshire to Dorset, and has 'Jurassic Coasts' at each end. Well known now that it has achieved World Heritage status, the now formally named Jurassic Coast exposes some magnificent geology between Exmouth and Purbeck. The northern version is rather shorter and has a reduced stratigraphic sequence, but reveals many of the same fossil-rich beds along a glorious stretch of coast around Whitby and Scarborough.

From Saltburn southwards, the eastern edge of the North York Moors has been eroded by the sea to form a spectacular coastline. Within the nearly horizontal sequence of interbedded sandstones and mudstones there is no thick unit of the stronger sandstone that can stand as a great wall. Consequently, the Moors' sea cliffs appear as broken slopes of crags and

Robin Hood's Bay with the older, lower part of the village now protected by its massive piled wall, and with low tide exposing the foreshore, ribbed with harder beds within the Redcar Mudstone.

ramparts, though they do reach heights of a little over 200 metres just west of Boulby. This makes for a beautifully varied coast where each headland has its own distinctive profile, each dictated by the geology.

A cliff-top walk can be especially rewarding at times of low tide, when wave-cut platforms are exposed along much of the Moors' coast. Planed off by wave erosion at high tide, some of these are scored by swirling patterns of rock ribs created by thin beds of harder stone that are exposed across very gentle fold structures.

Between platform and cliff-top, some sections of cliff are broken by benches of uneven ground known as undercliffs. These are rotational landslides formed where blocks of cliff have moved downwards and outwards over curved slip surfaces after slope toes have been removed by wave erosion. They are perfectly normal features of evolving coasts. South of Ravenscar an undercliff has formed where alternating strong and weak beds are especially prone to small landslides; it extends for three kilometres and carries footpaths that are low-level alternatives to the main cliff-top path. Around Runswick Bay, sections of undercliff within the Whitby Mudstones include individual landslide blocks prominent enough to warrant their own names: Seaveybog Hill lies round the corner to the east, and Wrack Hills lie north of the bay.

Undercliffs generally involve large, slow, creeping movements. Very different are small rockfalls, many of single blocks of stone, which are instantaneous and unpredictable events on steep cliffs of strong but fractured rock. It is fortunate that most occur after heavy rain or storm events, when there are fewer people on any beaches below.

Different again are mudslides from coastal cliffs of glacial till or weathered mudstone. These, too, are mainly triggered by rainfall events, which weaken the sub-surface by raising groundwater pore pressures, though movements typically take a few days or more.

The little bay at Port Mulgrave is backed by a steep bank that is a mass of old slides of saturated mudstone descending on to the beach. Fresh slides occurred in both 2018 and 2021, but slipped debris is quickly removed from the beach by wave action, and new plant cover soon makes the smaller landslides difficult to identify.

Bays and headlands

Between the unobtrusive headlands, bays form only gentle indentations along the Moors coastline. However, the main bays are large enough to trap sediment being moved along the coast by wave action, so they contain some lovely beaches, with pale quartz sand derived from erosion of the sandstone cliffs. The long, sandy beach between Redcar and Saltburn is not really part of the North York Moors, which start at Worsett Hill, capped by sandstone high above cliffs of Cleveland Ironstone (that is interbedded with mudstone and sandstone). Southwards from there, the cliffs and headlands vary with the detail of their local geology, and are broken by a succession of bays, each one different from its neighbour.

It is hardly a bay that features Cattersty Sands along the shores of the industrial landscape of Victorian ironstone mining at Skinningrove. Little of this terrain remains in its natural state, especially west of the beck where the slopes below the ironworks are largely old

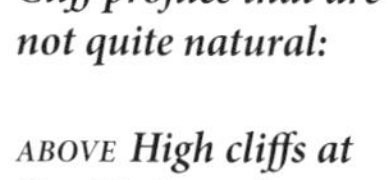

Cliff profiles that are not quite natural:

ABOVE ***High cliffs at Boubly bear the scars of quarries for alum shale; but sandstone cliffs in the distance, at Staithes, are in their natural state.***

LEFT ***The small bay at Skinningrove has its slopes beyond greatly modified by banks of waste from the old ironstone mines.***

Rias in Miniature

Whitby stands astride a lovely ria, which has given it the shelter of an excellent harbour, and is one of a select few along the moorland coast. A ria is a drowned valley, excavated by a river when sea level was lower than it is today, and then inundated by a rising sea level. Low sea levels were a feature of glacial stages, when worldwide sea level fell by more than 100 metres while so much water was held in huge continental ice sheets. So the greatest rias were formed along coastlines that were not covered in ice but were breached by active rivers: Milford Haven, Plymouth Sound and Salcombe Harbour are fine examples, all south of the ice during most of the glacial stages.

Yorkshire's coast was under North Sea ice during most of both the Anglian and Devensian glaciations. So the Moors' rivers could carve their valleys to greater depths only during shorter intervals of time, when ice margins had retreated while sea levels were still down due to water held in the remnants of the ice sheets, or during other cold stages of the Quaternary when the Scandinavian–Scottish Ice-sheet never fully developed but sea levels were somewhat lowered.

During any or many of these intervals, the River Esk carved its valley far below present sea level, and post-glacial submergence created the ria that became the natural harbour (above) where Whitby's fishing fleet was moored until recent times, and from where James Cook first set out to sea in 1746. The erstwhile captain had previously lived for a time in the fishing village of Staithes, which is built around its own lovely ria (below), albeit even smaller than the one at Whitby – in fact it is perhaps not a true ria as it is little more than a bouncing stream at low tide. Smaller still is Boggle Hole, cut into the cliffs of Robin Hood's Bay, though this cannot count as a ria either because it is left high and dry at low tide.

waste tips, now grassed over. The miners' long concrete jetty has had its own impact, trapping longshore drift on its western side to maintain a wide beach, whereas there is not much sand down-drift around the outlet of Kilton Beck.

Further south, the first properly indented bay is Runswick, ringed by low cliffs of Whitby Mudstone with a cap of Ravenscar sandstones that keeps faces in the exposed mudstone over-steepened and rather unstable. The bay appears to have formed where two valleys entering from the south had breached the protective cap of sandstone so that cliff retreat was more rapid due to wave erosion and slumping of the undercut mudstones. A village clings to the mudstone cliff at the western end of the bay, but a smaller ancestor, immediately to the north, lost nearly all its houses atop a landslide into the sea in 1682. At the eastern end of the bay, the Kettleness headland has lost its prominence to a large quarry that yielded alum shale until 1861. The site also had a small miners' village that was almost entirely destroyed by a landslide in 1829, though this ground failure was due more to expansion of the quarry than to erosion by the sea.

The wide sweep of Sandsend Bay has expanded eastwards from its pre-glacial ancestor that is now occupied by the two valleys containing Mulgrave Woods, with the castle ruins atop the low intervening ridge. That ancestral feature, whether one or a pair of valleys, was filled with glacial till that was scoured by wave erosion more readily than the bedrock of mudstones and sandstones now so well exposed in the cliffs at Whitby.

The north side of Robin Hood's Bay is a fairly typical part of the Moors coastline, with an eroding cliff of crumbly sandstones overlying weaker mudstones.

Those Ravenscar sandstones underlie the great bench on which stand the remains of Whitby Abbey, and form the magnificent prow that overlooks the eastern side of Whitby harbour. However, even strong rocks eventually succumb to relentless erosion by the sea, and these face the full force of waves and weather from the North Sea. The Haggerlythe is a rock shelf overlooking the harbour and backed by the high sandstone cliff. Houses stood on the shelf for many years, until they were destroyed by a great landslide in 1786; it may have looked like a prime location, but the strong sandstones are underlain by weak mudstone, and the entire profile of cliff and shelf is essentially unstable. Smaller landslides have followed the big event, but houses are no

Whitby with the ruins of its abbey atop sandstones of the East Cliff overlooking the drowned valley, or ria, that has become a sheltered natural harbour.

longer situated on the exposed end. Whitby's West Cliff is different, because a north–south fault underlies the harbour, west of which the sandstones extend above and below sea level, creating a stable cliff profile capped by slopes in the thick glacial till.

Robin Hood's Bay is cut entirely in weak Redcar Mudstone, and has very little in the way of a decent beach. Its cliffs yield minimal sand of their own, and the bay is not indented enough to trap sand arriving by longshore drift; any coming in from the north is soon swept out of the southern end of the bay. Instead, low tide reveals a magnificent wave-cut platform that, along with the low-fringing cliffs, is famously fossiliferous, and especially noted for some of its splendid ammonites. These Lower Lias beds are brought up to sea level, and consequent wave erosion, by the very gentle Cleveland Anticline.

Headlands at each end of the bay, and most of the cliffs north of the village, are formed of the stronger Staithes Sandstone. But the village, also named Robin Hood's Bay, has been built ridiculously close to the lip of weak cliffs of weathered mudstone and glacial till, much of which is actually a ramp of ancient landslide debris. Over the centuries it has lost more than a hundred houses to advancing coast erosion, and is now protected by a massive wall of concrete, up to 29 metres tall, which was built in 1975 and considerably enhanced by later works. In 1790 a landslide destroyed the village's approach road along with two lines of cottages; its site is now recognisable as the great grassy bank that carries the new road and upper car park on its rim and a barrier of armour stone at its foot, just north of the concrete wall.

Scarborough has been a holiday town since its mineralised spa waters were first exploited during the 1600s, and it grew into a major resort when the railway arrived in Victorian times. This perfect holiday town has wide beaches in two grand bays that are separated by the dramatic headland of Castle Rock. The headland is a down-faulted block of strong Calcareous Grit separated from the mainland by faults that lie parallel to the coast. It is an isolated remnant of a slice of the Grit that was originally more extensive at and just above sea level. To north and south this has been lost to wave erosion, which subsequently progressed across the faults and cut Scarborough's two bays back into the sandstones and mudstones that are weaker than the headland's grit. Wrapped around the headland, Marine Drive stands on a substantial sea defence that was completed in 1907; this has proved worthwhile because the cliff of strong grit is underlain by weak Oxford Clay that had been suffering from wave action.

The headland capped by Scarborough Castle is a block of calcareous grit, separated from the mainland by a fault; its Marine Drive stands on rock armour that prevents erosion of the weak clay underlying the strong grit.

The sweep of Scarborough's South Bay towards the Castle headland, a block of strong sandstones that is isolated beyond a fault hidden beneath the Bay sands.

Coastal landslides

Whether old or not so old, landslides are numerous along the Yorkshire coast. They are, after all, just one end of the spectrum of processes that ensure degradation of any steep slope, and particularly any undercut by wave action. The undercliffs that are a repeating feature along the Moors coastline of Jurassic rocks have been the site of many destructive landslides. In 1757, the alum works built on the undercliff at Loftus were destroyed by a landslide, and then in 1858 the kilns and furnaces in front of the ironstone mines at Wrack Hills were destroyed when the undercliff on which they stood slumped towards the sea.

There are no undercliffs at Scarborough, but South Cliff, backing the South Bay, is an almost continuous chain of prehistoric and recent landslides developed within the glacial till, with bedrock only exposed for a short way above beach level towards the southern end of the bay. The spectacular landslide at Holbeck Hall in 1993 (*see* page 94) is still a recognisable feature, and to the north of it, two more landslides have moved in recent times.

The Clock Café stands on the lower part of a landslide that crept downhill a little more than usual in 2018. This cracked a footpath and pushed a retaining wall on to the historic beach huts that stood on the toe of the landslide; they were subsequently demolished, but were scheduled for rebuilding in 2023.

Further north, The Spa was based on springs emanating from glacial till, though the water, rich in magnesium sulphate (Epsom salt) and iron carbonate, is likely derived from buried Cornbrash limestone and Ravenscar sandstone. The combination of healthy spa waters and a coastal location established Scarborough as one of Britain's first seaside resorts in the late 1600s. The Spa centre was destroyed by a great landslide in 1737; it was rebuilt on the same site, but was then destroyed by fire – the present buildings date from 1879. 'Taking the waters' has lost its popularity, so The Spa has evolved into an entertainment complex. But it is still backed by a high slope of inherently unstable glacial till; since 2019 this has been stabilised with an array of ground anchors and geogrid, now hidden beneath the soil and grass.

The small concrete 'headland' below the Italian Gardens replaces the seawater lido that was closed in 1989, and the bowl of grass slopes above it is the scar of a older landslide. The 1993 landslide at Holbeck Hall was the big event within South Bay. Farther south, many small mudslides are still active within the slope below the golf course and above the low bedrock cliff. This coastline continues to be geologically active, though parts have been tamed by grand engineering works

Osgodby Point at the northern end of Cayton Bay, backed by the tree-covered bowl that is the slow-moving Cayton Cliff landslide; the houses on the skyline survived when four others were lost to the landslide in 2008.

Filey's coastline

Between Moors and Wolds lies the Vale of Pickering, and its eastern end also meets the coast, or very nearly so. But this section of coast, on either side of Filey, bears little relationship to the Vale. In place of the Vale's lake sediments on a clay floor, the coast has a rib of Jurassic bedrock only just exposed from beneath a barrier of glacial till.

From White Nab at the southern end of Scarborough's South Bay to Filey Brigg, the line of cliffs is largely formed by a narrow extension of the strong rocks of the Tabular Hills, notably the Lower Calcareous Grit, all dipping gently inland. Cayton Bay is the exception, cut back into the Oxford Clay and a mixed sequence beneath. At the northern end of the bay, and west of a small fault, the clay forms the entire cliff, from a height of 50 metres to below beach level. So in place of any real cliff, there is the large, deep-seated Cayton Cliff landslide. Its most recent significant movement was in 2008, when four houses were lost on its crest; the main coast road was also threatened, so has since been rebuilt further inland. South of Cayton Bay, Lower Calcareous Grit stands above, and is undercut by, the Oxford Clay to form a rocky coastline, culminating in the great wall of Newbiggin Cliff.

The high headland of Filey Brigg is formed largely of glacial till, but this survives because it stands on a narrow, sloping platform of the strong Calcareous Grit. On the exposed northern side, the grit reaches to twenty metres above sea level, keeping the till well clear of wave attack,

Filey Bay is cut back into the coast on the outcrop of the soft Kimmeridge Clay, though the town stands high on glacial till that is effectively a very large lateral moraine left by the North Sea Ice Sheet.

Holbeck Hall Landslide

Scarborough grabbed people's attention in the summer of 1993 when a spectacular landslide on the South Bay developed slowly enough to be watched as it progressively destroyed the Holbeck Hall Hotel; this building had stood there for a hundred years, nearly 60 metres above the beach. On 4 June, residents went down to breakfast to find that the gardens in front of the dining room had disappeared overnight. Glacial till underlies the site, and a rotational landslide had developed through its entire thickness. Fortunately there was time to evacuate the hotel before the inevitable regression of the new, steep head scar saw multiple smaller landslides adding to the sliding mass of debris that extended as a flow out on to the foreshore. Close to a million tonnes of glacial till was on the move, and a couple of days later, the entire dining room crumbled on to the head of the landslide.

This was a classic rotational landslide over a curved slip surface, and was not a feature of any current marine erosion. A low sea cliff in the underlying Jurassic sandstone, and the modest concrete sea wall, were untouched by the landslide, except that the debris flowed over them.

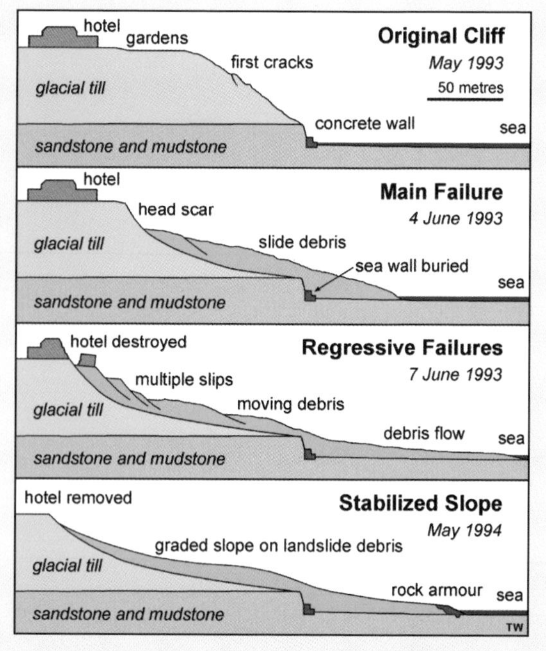

ABOVE *Stages in development of the landslide that destroyed Scarborough's Holbeck Hall Hotel in 1993, from the initial cracks, to the main rotational failure, followed by headward regression, and then stabilised by engineering works.*

BELOW *The landslide on Scarborough's South Cliff in 1993, slowly eating back beneath the crumbling hotel.*

Landslide in action at Scarborough in 1993, when the dining room block of the Holbeck Hall Hotel was at a jaunty angle, not long before its total collapse as the landslide extended headwards.

Failure of the slope was preceded by a long spell of dry weather that opened shrinkage fractures in the clay-rich till. This was followed by heavy rain that raised pore-water pressures in the ground, the effects of which were probably exacerbated by local construction work that disturbed the natural drainage. Less than a week of spectacular activity had been preceded by a month of tiny movements, which had not then been seen as significant.

To prevent further regressive failures, subsequent remedial works cut back the head scar, and thereby also demolished the remains of the hotel. Natural wave erosion would eventually have removed the debris fan on the foreshore, but this was left in place and protected with a rampart of five-tonne blocks of stone, thereby creating a more stable toe on the slope. The lower part of the landslide is now a ramp of grassy parkland, crossed by footpaths that include the Cleveland Way. Signs of the events of 1993 survive in the fenced-off, broken ground further up the slope; uneven steps are degraded head scars of subsidiary slips within the landslide mass, and small fresh scars indicate that the slide debris is still settling against its new toe of stone blocks. Except for its new sea defences, the site has now almost blended into the coastal scenery.

The grassy bank midway along Scarborough's South Cliff is the debris of the 1993 landslide, with the arc of the new bank of armour stome that was built round its toe to prevent fiurther erosion and sliding.

whereas the till descends almost to beach level against the calmer waters of Filey Bay. The outer half of the Brigg is a rib of grit exposed only at low tide and stripped of any till cover.

Filey Bay exists where the coast encounters the lowland formed on two weak clays – the Jurassic Kimmeridge and the Cretaceous Speeton. However, neither of these clays has any coastal outcrop, because they now lie only below sea level. The coast is a rampart of glacial till, the lateral moraine left by Devensian North Sea Ice. Rising 20–40 metres above the long sandy beaches of Filey, Hunmanby and Reighton, the till is draped over Jurassic rocks northwards past Scarborough and over the Cretaceous chalk south of the Speeton Hills – but at Filey Bay it is all there is to keep the sea out of the Vale of Pickering. For practically its whole length, this bank of till down to the beach is a succession of shallow landslides repeatedly active as the sea cuts into their toes.

A marine arch of chalk within the Bempton Cliffs.

Chalk cliffs of Flamborough

Marked by the front of the Speeton Hills, a small fault brings Filey Bay to an abrupt end where the largely vertical chalk cliffs rise to rhe south. Chalk is not a strong rock, yet it forms beautiful white cliffs. This is because it generally lacks inclined planar fractures that could destabilise its rock mass. Wave action undercuts the chalk, and the rock above falls away in small blocks or thin slabs, thereby creating cliffs that are close to vertical. Small rockfalls are common from these chalk cliffs, but come from the upper and lower parts in similar amounts; so the cliffs do retreat, albeit very slowly, and normally retain their vertical profiles. The Bempton Cliffs are up to 100 metres tall, with vertical bare chalk capped by a low grassy ramp on glacial till that is only a few metres thick. Spectacular indeed, but they are famed most for the thousands of sea birds that nest on the tiniest of ledges.

Robin Lythe's Cave, in the chalk headland on the east side of North Landing at Flamborough Head. The main entrance from the open sea is behind the camera, and the narrow entrance from the bay is to the right of the person standing near the back wall.

The cliffs of Flamborough Head, with vertical walls of white chalk capped by grassy slopes in the cover of glacial till, looking north across Selwicks Bay.

The great chalk cliffs continue southwards to Flamborough Head, but decline in height to only about 30 metres of vertical chalk, mostly capped by around twenty metres of glacial till. The latter has slumped into debris flows that have tipped over the cliffs, so that flat land is typically set back 30 or 50 metres from the lip of the vertical chalk. The Flamborough coast is indented with narrow inlets and a few wider bays, unlike the straight line at Bempton. Wave action is very selective, and rapidly eats into any weakness in the rock. So the cliffs are riddled with sea caves, some of which reach in nearly 100 metres. Roof breakdown has turned others into long narrow inlets, known as geos. A few caves reach through necks of land to create fine marine arches, and collapse of their roofs has left some classic sea stacks. The arches, geos and stacks are spectacular all along the northern side of Flamborough Head; this is where selective erosion is greatest under the impact of storm waves coming in from the north – the southern side of the headland is more sheltered, and its cliffs follow a smoother line. Low tide reveals wide, wave-cut platforms in the small bays, and the whole headland provides a splendid seascape of marine erosion at work.

A marine arch in the chalk cliffs on the west side of Flamborough's North Landing, destined to become a sea stack when the weathered chalk forming the arch roof inevitably collapses.

The splendid sea stack of nearly horizontal chalk on the south side of Selwicks Bay.

The buried Ipswichian cliff at Sewerby, where the present chalk cliffs (capped by glacial till and a thin layer of cover gravels) end against the grass-covered bank of till that extends unseen below beach level.

Further south, the chalk cliffs end abruptly below the village of Sewerby, to be replaced southwards by the low cliffs of Holderness till. The break at Sewerby is where the pre-glacial chalk cliff heads inland, buried by the glacial debris. The pre-glacial cliff is aligned close to east–west, and the modern coast aligns closer to north–south, so the new cliff provides an oblique cross-section through the old cliff. Sadly the details are rarely visible, because the chalk cliff is less than 15 metres tall beneath nearly 20 metres of glacial till; the latter breaks down into debris flows that cascade over the chalk, and the extent of clean rock varies from year to year. However, the site is regarded as internationally important because the beach deposit at the foot of the buried cliff dates to 132,000 years ago, when the interglacial sea level declined at the start of the Devensian glacial period.

Holderness

South of Sewerby, and reaching to the Humber Estuary, Holderness is a wide plain with minimal local relief. Much of it was wetland until it was drained to provide useful farmland, and Hornsea Mere is a remnant of extensive patches of bygone marsh and open water.

Prior to the Last Glaciation, Holderness did not exist. Across most of it, bedrock chalk is below sea level, and the pre-glacial shoreline south of Bridlington formed a wide sweep past Great Driffield and west of Beverley and Hull. Then the wasting fringes of the Devensian North Sea Ice-sheet dumped up to 50 metres of glacial till. When post-glacial sea level returned to near its current position, a new coastline stood at least 10 km out from the present shores. But a land of soft, weak glacial till has been no

Coastal erosion in action on Holderness, between Skipsea and Ulrome when the coast road was falling away during 2009; since then, it has all gone.

match for relentless wave erosion by the North Sea. Furthermore, debris from the crumbling shoreline is soon carried away by longshore drift. Holderness tills are about one third sand, which provides sand and coarser sediment for the beaches, while silt and clay are carried away in suspension by waves and currents.

The direction of coastal sediment transport, known as longshore drift, is determined by the powerful waves that have travelled the greatest distance. Waves wash sand obliquely up a beach but the returning swash takes it directly down the slope. So waves from one side shift sand zig-zag along a beach. In the North Sea, the big waves come from the north, so beach sediment, including debris from the failing cliffs, moves to the south, and it moves very rapidly on a coast as exposed as Holderness. Then, once gone, the land is subjected to further wave attack.

The result is coastal retreat on a grand scale, on average by somewhere between one and two metres per year. Records reaching back to Roman times tell of more than two kilometres of retreat along the entire Holderness coast, and at least 30 villages have been lost to the waves. The present coast is a line of cliffs, mostly ten to twenty metres tall. From Bridlington to just short of Spurn Head they are all cut in the glacial till left when ice melted away after the Last Glacial Maximum. This clay-rich material stands in almost vertical faces – for a short time – until it is undercut by wave erosion. Sandy beaches are covered at high tide, when waves eat into the weak till, until it fails in repeated small landslides.

A typical landslide removes a slice of cliff, cutting the rim back by a few metres along a length of 30–100 metres. Movement is rotational, so blocks of ground slump down and outwards, atop slip surfaces that emerge close to beach level. Some sites have multiple slices stepping down the cliff, but all the fallen material is eventually removed by wave action, so that, all along the coast, the process repeats every few years, commonly after winter storms have eaten into the base of the cliffs.

These landslides are classic features of the Holderness coast; there are always some to be seen, but the situation changes almost every year. Various figures have been calculated for a current, long-term, average rate of coastal retreat on Holderness, but most range between 1.5 and 2.5 metres per year.

A section of coast directly east from the village of Aldbrough has a cliff of glacial till that rises about 12 metres above the beach, and is retreating at a rate of about 2.4 metres per year. That is the mean rate, as land is lost in bites that remove a few metres at a time. An entire Victorian holiday village, with its own hotels

Repeated landslides in the glacial till due to coast erosion at Holmpton, towards the southern end of Holderness.

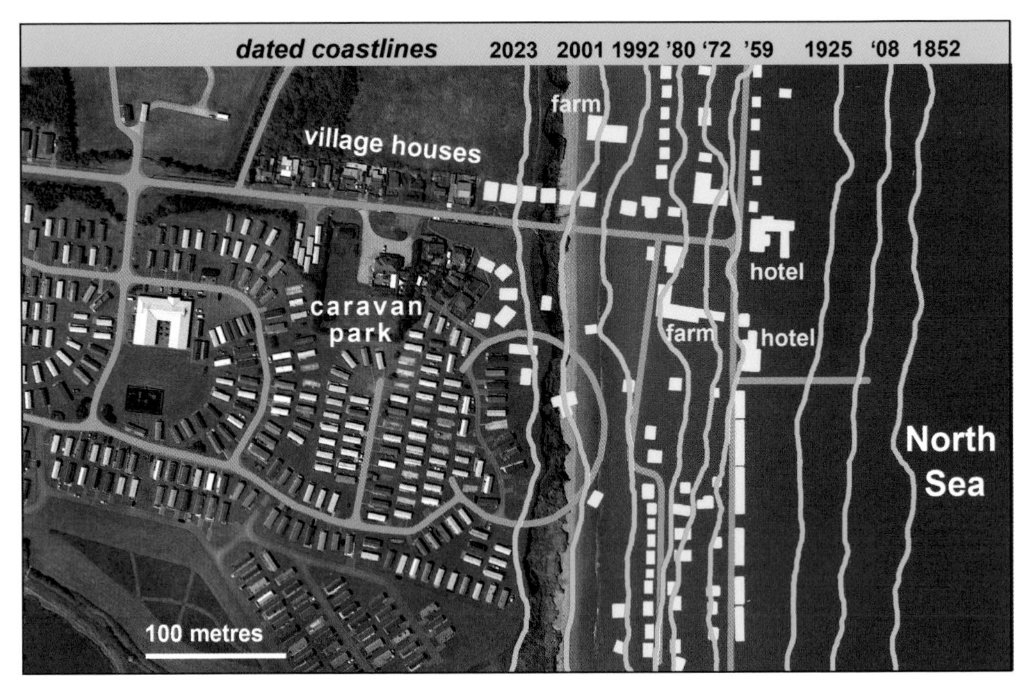

Long-term record of coastal retreat at Aldbrough on the Holderness coast. The satellite image dates from 2009, and many of the caravans have been moved inland since then. The blue dated coastlines are derived from old maps, air photographs and ground surveys. Buildings marked in yellow have either fallen into the sea or were demolished prior to their demise.

near the coast, was lost to the sea in the early 1900s. Currently the coastal land is occupied by a large caravan park, which is the ideal use for the site, as the caravans (now mostly large, and better described as trailer homes) can simply be rolled back as the sea slowly chews into the land; an adjacent line of smaller houses consists mainly of timber structures with a limited lifespan.

Aldbrough has become a classic site where the repeated landslides and coastal retreat have been recorded and monitored over the years. In the short term, retreat rates increase slightly in years of higher rainfall or more frequent storms. Periodic variations are produced by zones of low beach level, known as ords. Separated by banks of sediment where the beach is a metre or so higher, these migrate along the coast at annual rates of around 500 metres, and cliff erosion is increased when it overlooks the trough of an ord, where there is therefore less protection by the beach. There are still details of the coast erosion that are not fully understood, but the overall picture of unrelenting cliff retreat is very clear.

One of the farmhouses at Aldbrough collapsing into the sea during the winter of 1980, beside the remains of the road that had once extended out to the seaside hotels.

Stabilising an eroding coast

All along the Holderness coast, land is eroded away at an alarming rate. Villages, farms, homes, holiday cottages, roads and farmland are all being lost to the sea. Prevention of this with a full-length sea wall is unrealistic, impractical, fraught with side issues and economically unviable. So a compromise solution has been devised, and is now in progress. It all depends on the concept of eventual coastal stability between 'hard points' that are protected and maintained to be erosion proof; intervening bays then deepen until they trap enough beach sand to achieve some measure of equilibrium. Each hard point should evolve with a parabolic bay on its down-drift side, which then becomes stable, all for the cost of erosion defences only on the hard point.

Flamborough Head is a natural hard point. Though chalk is hardly a strong rock, its great mass at the eastern end of the Wolds is slow to erode, and it therefore stands proud as a very grand headland. The entire Holderness coast is its down-drift parabolic bay. This has been deepening ever since the glacial retreat, but, left to nature, the sea would advance by eroding through many more kilometres of land before the coastline reached equilibrium and thereby stabilised.

ABOVE Blocks of gneiss that were imported from quarries on the south coast of Norway, and now a bank of armour stone placed to protect the Holderness coastline at Mappleton.

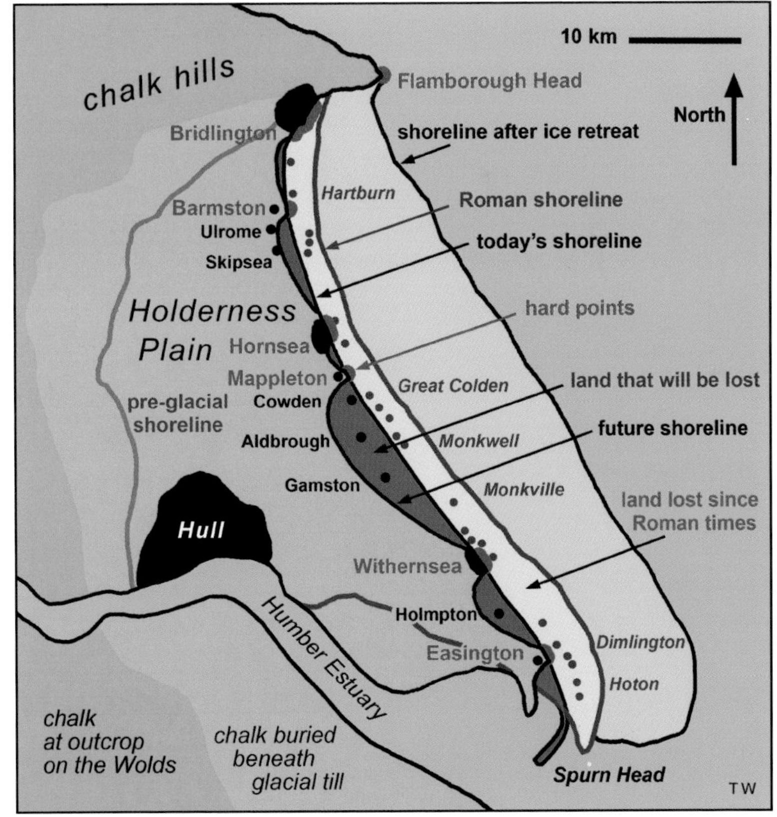

RIGHT Notable features of the Holderness coast, with some of its many historical shorelines and positions of the lost villages; this also shows the master plan that is in progress with hard points created to establish a coast that should become stable.

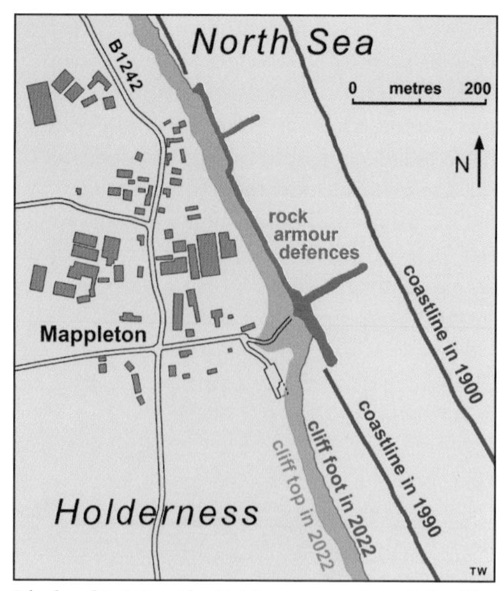

The hard point on the Holderness coast created with rock armour at Mappleton, where coastal retreat has ceased at the village but has continued in the stretch that is down-drift to the south.

So the modern coastal defences have five artificial hard points, at Barmston, Hornsea, Mappleton, Withernsea and Easington, with a sixth at Bridlington that has long been part of the town's structure. Together these are starting to develop a crenulated coastline with a reduced rate of overall land loss. Hornsea and Withernsea are towns with long histories of coastal defences; the former had its grand sea wall built in 1930. Easington is protected because its huge gas terminal is fed by pipelines from the North Sea gasfields. Barmston is a smaller and older version of the most recent defences that were built at Mappleton in 1991.

Mappleton is a village no larger than many others, but it gained its coastal defences because it is fortuitously located where the 'strategically important' B1242 coast road was in imminent danger of destruction. Adding to some smaller, older defences, the new structure consists of two beach groynes forming an L-shape. These are made of five-tonne blocks of Norwegian gneiss, each too large to be moved by wave action. The stone came from afar because sea transport is very cheap, and barges were easily beached at high tide and then unloaded by front-loader trucks at low tide; the blocks now provide an unexpected delight of metamorphic geology.

Hard points resist any erosion by the sea, and also trap some of the longshore drift to create a protective beach, while inevitably deflecting some sediment out into deeper water. They therefore starve their immediate down-drift coastline of its regular supply of beach sand, and the automatic consequence is increased cliff erosion to replace the lost sand input. Coastal retreat increases as a bay deepens in its aim for equilibrium in its sediment transport. The grand plan for Holderness is to end up with five new bays. It is estimated that the smallest

Relentless wave attack has caused the unprotected coast to continue to retreat where it is down-drift of the bank of armour stone (with its end visible on the far left) at Mappleton, now one of the hard points designed to reduce the loss of land on Holderness.

Four years of coastal erosion result in the loss of houses that were not protected by the stone defences at Ulrome, Holderness.

bay, down-drift of Hornsea, will stabilise within 200 years or so, whereas the largest bay, down-drift of Mappleton, will take 2000 years before it reaches a fragile equilibrium.

In the face of relentless marine erosion, this system of hard points is the best option. However, it does mean that some parts of the coast are subject to more rapid retreat than in the natural situation without no hard points. At Great Cowden, a kilometre down-drift of Mappleton, mean annual cliff retreat increased from 3.0 to 4.7 metres after placement of new defences at the latter site. Farms at Great Cowden therefore lost their land more rapidly than they could have expected on an undisturbed coastline. It is bizarre that such farms receive no compensation for the years of lost production, while other sites benefit from creation of the hard points. Much of this inequality derives from a tragic decision at a Lands Tribunal in 1998 made after the clear facts of coast erosion were subsumed by legal trivia. Holderness will long remain a scene of 'geology in action' that creates controversy and difficulty for those that live and work there.

Spurn Head

The last chance for beach sediment southbound along the Holderness coast is Spurn Head. With anything swept beyond lost into the currents of the Humber Estuary, Spurn is a major zone of deposition. It is a fine example of a sand spit, and furthermore it is shaped into a hooked spit by wave refraction that curves round its end. Built with material swept from the Holderness coast, it is topped by ephemeral sand dunes, but it is all rather too thin and fragile in the face of North Sea storm waves.

Probably deriving its name from a spur, or projection, Spurn Head is a blob of land just 300 metres wide and twice as long; it is connected to Holderness by a strip of sand nearly five kilometres long and only a few tens of metres wide. A large spit has existed at the site since AD600 and probably well before that. Its head was occupied by the medieval port of Ravenser Odd, which was then more important than Hull, until it was destroyed by a storm in 1362, just five years after it had been abandoned in the face of a previous storm. Now, Spurn Head has an automatic lighthouse. For many

Built of sediment eroded from Holderness, the long sand spit at Spurn reaches into the Humber estuary.

Land of glacial till at Kilnsea stands only a few metres above sea level on the southern tip Holderness; concrete slabs on the beach once supported chalets, and the sand continues behind the camera for 6 km to Spurn Head.

years it had the Humber Lifeboat Station, with the only full-time lifeboat crews in Britain as there was no local employment to support the traditional volunteers, but in 2023, lifeboat and crews were transferred to Grimsby.

The landward end of the Spurn spit migrates westwards to stay attached to the retreating coastline of Holderness. The outer parts of the spit are also shifting to the west, but probably more slowly; the loss of Ravenser Odd was one consequence of this steady westward shift. Some confusion over early records led to an early concept of the Spurn spit being a cyclical feature, destroyed by storms about every 250 years or so, then rebuilt by the sea at a new position further west. This is now refuted and the westward migration is considered to have been almost steady since about 1700, albeit very slow.

Longshore drift carries sand southwards along the spit's outer shore, and waves are refracted enough to carry some of it round the end to add to the intertidal flats. But the variable movement of sand can leave the northern part too thin and too low. In 1849 a great storm breached the spit and left a gap that a small boat could be sailed through; there may have been earlier breaches, but reports are unreliable. That 1849 loss was probably due in part to nearby extraction of beach sand, a detrimental activity that ceased thereafter. The breach was filled with blocks of chalk, groynes were built to stabilise the beach, and marram grass was planted to hold the sand dunes. But these features were not maintained, so the site has reverted to more natural processes. A storm in 2013 failed to breach the spit, but wave overwash carried away the roadbed and enough sand that 100 metres of the crest was submerged during higher tides. However, coastal evolution never ceases, and new sand dunes are expanding in a belt west of their ancestors; the new strip of land is less frequently submerged by high tides.

Clearly, the entire Spurn sand spit is a fragile feature, and further changes are inevitable. But these are rather unpredictable, because the site lies in the wake of the recent protection works along the Holderness coast; so the past is little guide to the future. Within the erosional environment of an exposed coast, its long-term survival cannot be assured, but for now at least, Spurn Head is a magnificent and distinctive feature of the Yorkshire coast.

Blown sand on the eroded end of the road to Spurn Head, with the low sand spit extending into the right distance.

CHAPTER 9

Minerals and Mines

Since time immemorial mankind has exploited the ground beneath his feet. Coal, iron and salt, along with stone to build houses, were among the first materials to be sourced from the bedrock, and people were therefore beholden to geology before they realised it. Then, as society steadily evolved in complexity, the sheer scale of quarries and mines increased at the key sites; this was followed by an inevitable decline after efficient transport allowed raw materials to be sourced globally instead of locally.

Eastern Yorkshire has been no exception, particularly where the sedimentary rocks of the Moors and Cleveland Hills have yielded a host of materials to profitable extraction. They provided massive wealth in the past, before declining to a mere handful of operations today.

A clearing in woodland with a grassed-over mound of rock debris. Such are the typical remains of a bygone mine with a drift entrance now blocked by soil and lost in the trees. This was one of many small mines in alum shale near Littlebeck, south of Eskdale.

In times long gone, gravel, sand and clay were extracted from numerous pits to supply the construction industry, though none reached major proportions, and most have since blended into the landscape. Lime was also produced by working the chalk at many sites across its Wolds outcrop. A few of these have continued to produce raw material for the cement industry, creating some very large chalk quarries in the southern Wolds. Scattered across the hills, quarries in the stronger sandstones and limestones tended to leave rock scars, many of which are now crags that are welcome features of interest within the landscape; now only a few quarries remain active. In the past these were the local sources of stone for all kinds of building.

Chalk exposed within a large modern quarry cut into the far southern end of the Wolds.

Whereas building materials can be garnered from a host of dispersed sites, extraction of more valuable minerals has always been restricted to specific locations, dictated by the geology. There is no gold or silver, but the North York Moors and Cleveland Hills have yielded minerals of great value, some in the past and others in the present. The upper half of the Lower Jurassic sequence, which used be known as the Lias, is remarkable in that it contains resources of three totally separate minerals that have been worked extensively in the past – alum, jet and ironstone. Subsequently potash minerals came to the fore, extracted from far beneath the Moors. And now, not to be left out, bits of Yorkshire south of the Moors are revealing hidden resources of natural gas.

Alum shale

From generating the first great industry of eastern Yorkshire, alum has long since shrunk back into obscurity. It had a complex history. In the Middle Ages, Britain's wealth was founded on textiles and natural fibres, and dyed cloth was a major element. This required a mordant, a chemical that creates a bond to hold a natural dye on to a fibre, so that it will not fade or wash out. Alum (a double sulphate of aluminium with potassium or ammonium) was that mordant, but it does not occur naturally in Britain. Huge amounts were imported, and by the mid-1500s nearly all of it was coming from the papal mines in volcanic rocks at Tolfa, north of Rome. Then the Church of England

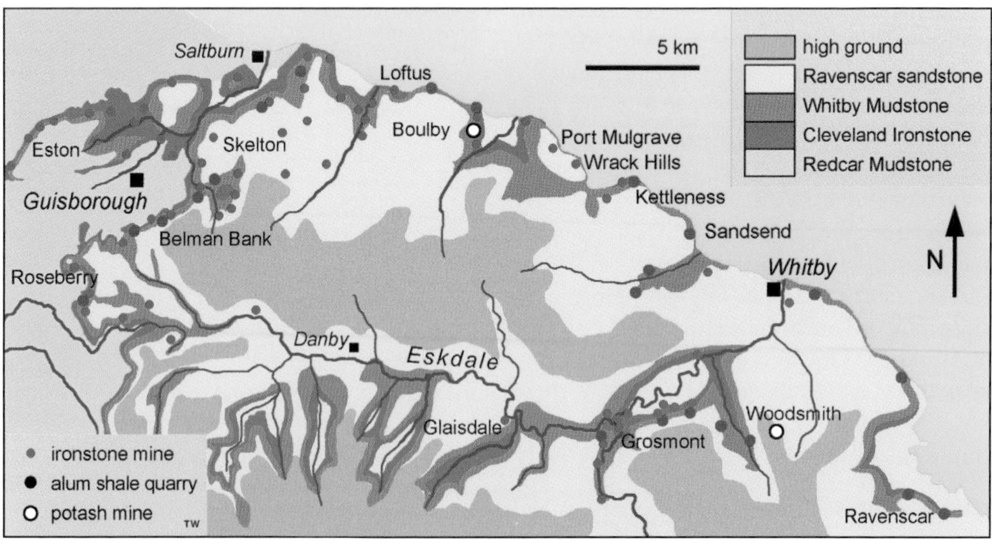

Mines and quarries of ironstone, alum shale and potash in the North York Moors and Cleveland Hills. The stratigraphic units are generalised and include beds of other lithologies. Named mines are referred to in the text.

Next to the main road east of Guisborough, the old Slapewath quarry in alum shale is now a woodland dotted with blocks of the sandstone caprock and laced with trails for motorcycle scrambling.

Remains of the alum works at Ravenscar, with the quarry that provided the shale in shadow below the skyline.

split from Catholicism, and exports from Italy ceased; Britain became desperate to produce its own alum, with many companies losing a lot of money on failed attempts to manufacture the precious mordant.

Thomas Challoner had been to the Tolfa mines in Italy, and was a keen botanist: he noticed that some of the plants growing on his estates at Guisborough were similar to those at Tolfa, and his lateral thinking discovered that one part of the Whitby Mudstone contained the right combination of clay minerals, pyrite and organic carbon – it is now known as the Alum Shale (*see* page 14). This does not contain alunite, the natural potash alum that is mined at Tolfa, but it had the critical amounts of aluminium and sulphur to yield alum by a long and complex process that took a year to complete.

Challoner experimented with a small quarry in the shale at Belman Bank, above Guisborough, then in 1604 built the first alum works with its own quarry at Slapewath, further east on his estate. From there, the world's first great chemical industry expanded around the North York Moors, producing alum from raw material in the Whitby Mudstone. The bedrock was hacked out from large quarries, and other components were brought in, to produce up to

The headland of Kettleness, completely transformed by the quarrying of alum shales in years gone by.

5000 tonnes of alum per year at some 30 sites around the fringes of the Moors, mostly along the coast either side of Whitby.

Vast amounts of shale were quarried from hillside outcrops. Barrowed out on to the quarry floor, it was piled on top of a metre or so of brushwood and gorse, until it stood fifteen metres or more tall. The brushwood was set alight, and, fuelled by the carbon content of the shale, these piles could burn for up to a year. This allowed the clay minerals to break down and release their aluminium, which formed a sulphate by reaction with sulphuric acid being generated by oxidation of the wet pyrite. The burnt shale with its new sulphate content was then carted down to great open-air leaching tanks, each up to ten metres long, where the shale was steeped in water to extract the soluble sulphates of iron and alumina. The mineral solutions were passed from one tank to another in order to juggle their chemistry and allow impurities to settle out or float off. The purified solution was then allowed to flow along channels and into the alum house, where it was boiled to concentrate the sulphates, while the barren shale was tipped aside.

At that stage, an alkali was added to react with the aluminium sulphate to form alum. Burnt kelp was added to create potash alum, and human urine was added to create ammonium alum. Both varieties were mordants, and most Yorkshire works produced a mixed alum. Kelp was gathered locally, spread out to dry, and then burned to produce potash-rich ash known as wrack. Large quantities of human urine were gathered from afar – Londoners were even paid for their urine. It is slightly disturbing that the sea-borne transport maximised its use of wooden tubs by shipping them north full of product for the alum works, and shipping them south filled with Yorkshire butter.

Inside the alum house, the resultant liquid was again heated in large, lead-lined pans held over coal fires. While the water was steadily boiled off, the alum crystallised before the iron salts, which could then be pumped off in solution. Finally purified, the alum was allowed to cool and solidify before being dried and bagged for dispatch.

Within the growing industry, coastal sites were favoured; a quarry could stand above the alum works with a jetty below, so that movement of large tonnages of rock and liquids was aided by gravity. From the early 1600s onwards, the alum industry was massive along the coast of the Moors – but by 1871

The old Loftus alum shale quarries on the undercliff west of Boulby; the moonscape is created on debris of sterile shale left by the miners, littered with blocks of sandstone fallen from the cliffs of overlying Ravenscar beds.

A small mine entry at the side of the beck passing through Skinningrove Beck marks one of the earliest attempts at mining the Cleveland Ironstone.

everything had closed down. This was due to the invention in 1855 of faster and cheaper processes to make alum, using coal shale and ammonia, a by-product from coal gas; this was followed by the creation of aniline dyes that contained their own fixatives.

Whereas the earliest alum works lay inland around Guisborough, the major producers were on the coast between Saltburn and Ravenscar. Alum shales were extracted in such great quantities that the cliff profiles at Loftus, Kettleness, Sandsend and elsewhere were completely changed by the bare quarries and huge waste tips. The Loftus site, spread along the undercliff far below the Cleveland Way footpath, resembles a moonscape of barren ground; hardly a feature of geological time, it is nevertheless a rather grand component of the modern landscape. Further east, the remains of the Ravenscar Alum Works are well preserved, so visitors can visualise the operations of bygone years.

Cleveland Ironstone

For more than a thousand years scattered outcrops of iron-rich rock all over the North York Moors were dug out to feed small bloomery furnaces. These heated the ore over charcoal, reducing it to metallic iron that could be worked into tools in local forges. Perhaps the most remarkable sites were in Bilsdale and Ryedale, where thin beds of siderite in the Eller Beck Formation were worked by or for the monks of Rievaulx Abbey. This could have led to greater things, because the monks had developed primitive blast furnaces prior to their being disbanded in 1538 with the Dissolution of the Monasteries. Then in the late 1700s on the other side of the Moors increasing amounts of ironstone were dug out from the coastal cliffs north of Whitby, and shipped to iron foundries on Tyneside. These ironstones were in thin beds, many just as horizons of nodules, convenient for transport north by ship, but none worthy of large-scale mining.

At outcrop, most of Cleveland's ironstone looks rather like an ordinary sandstone. But that is the weathered rock, and it is generally dark grey or brown when freshly broken. Its density, about an eighth greater than most rocks, is barely recognisable when just hefting a hand specimen. Most of the ores consist of ooliths (spherical concretions about a millimetre in diameter) of green chamosite (a complex iron silicate that may be in the form known as berthierite) set in a matrix of pale brown siderite (iron carbonate), with varying amounts of other silicate minerals. It also contains rusty brown limonite (iron hydroxide), though most of that was formed by alteration or weathering

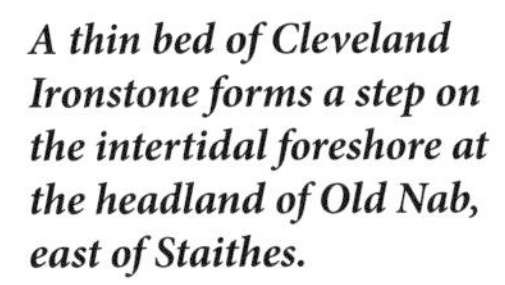

A thin bed of Cleveland Ironstone forms a step on the intertidal foreshore at the headland of Old Nab, east of Staithes.

of the primary iron minerals. Iron content varies from 32% down to 26%, and anything less is normally regarded as sub-economic and not practicable to extract. Rock only becomes ore when it can be mined at a profit.

When the railway was built in 1835 to link Pickering to Whitby, a short tunnel was required just south of Grosmont, and the excavations were found to run through a thick bed of ironstone. This initiated an iron industry of modest size in Eskdale – but better was yet to come.

John Vaughan had an iron works near Bishop Auckland, and had taken in some raw stone from the workings along the Moors coast. He realised that there should be more ironstone at inland sites, so set his mining engineer, John Marley, on a search along the northern slopes of the Cleveland Hills. In June 1850 the two Johns were prospecting along the hills above Eston, on the eastern edge of Middlesbrough. They found pebbles of ironstone, and then a small quarry that had been worked for roadstone, but was in fact cut into good iron ore. Those were the days, now long gone, that valuable mineral deposits could be found at outcrop simply because no one had looked for them previously. Others had found this iron ore, but it was John Vaughan who had the ability to exploit it. Within three months he had signed a deal with the landowner, dug quarries into the ironstone, built a tramway down to a new railway on the lowland, and delivered a first load to his own iron works.

From there the mining of Cleveland ironstone just grew and grew, and the statistics make impressive reading. By 1863 the local mines were supplying 78 blast furnaces on Teesside. From 1875 to 1914, some 8000 miners dug out around six million tonnes of iron ore each year. During the 1880s, Britain was producing about half the world's iron, and nearly 40% of that originated in the mines of Cleveland. But mineral deposits have finite extents, and the last of the region's mines ceased operations in 1964. By then, the iron industry on Teesside had become a steel industry, but that, too, was in decline. The last steel production (at Redcar, using imported ore) was in 2015, though steel is still milled at the Skinningrove site, which had first produced iron in 1874 when it took its raw ironstone from the nearby Loftus mines.

Mining of the Main Seam of the Cleveland Ironstone. From the outcrops the ironstone extends south beneath a cover of Whitby Mudstone and then far below the sandstones of the Moors. Its southern limit of working was determined largely by the declining iron content and a thickening middle of barren shale.

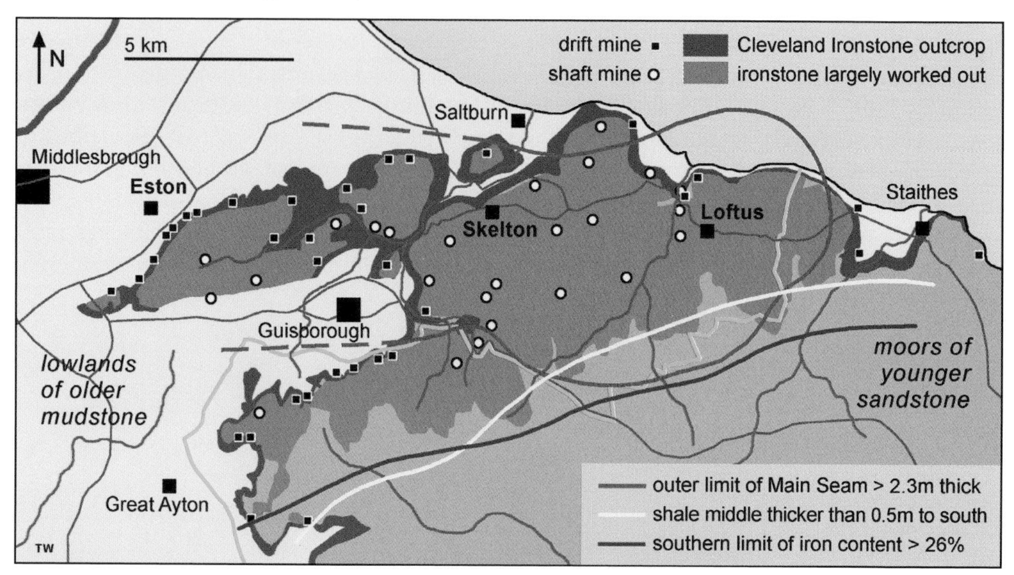

Drifts, shafts and retreat mining

Mining of the Cleveland ironstone has always adapted to local conditions. Being sedimentary, the ironstone occurred in multiple beds, or seams, with the Main Seam reaching up to nearly four metres thick. Other seams were typically around a metre thick where they could be mined. All seams were worked across limited areas, beyond which they are too thin, or split by too many 'middles' of barren shale, thereby rendering them sub-economic. Most significant was the southward deterioration of the Main Seam, where the stone contains less siderite so that its iron content is reduced; and this is paralleled by a shale middle becoming steadily thicker, until a combination of factors makes the seam impossible to mine while making a profit.

Lines of small quarries along outcrops were soon replaced by gently inclined tunnels, known as drifts or adits, following the seams, which dipped little below horizontal in most of Cleveland. These drift mines were followed by shaft mines where the ironstone was far from outcrop and lay as much as 200 metres below ground level. Sinking a shaft was an expensive operation that required a large outlay prior to any income from selling the mine product.

Within each seam, partial extraction of the ironstone was achieved by bord-and-pillar working, a local term for pillar-and-stall mining. Stone was extracted from galleries, or bords, driven on a grid pattern around pillars that were left to support the roof. With equal widths of bord and pillar, on a square grid, this could take out 75% of the ore. Most Cleveland mines had wider pillars, and so extracted only 30–50%. This was because they followed up with retreat mining, whereby most of the pillars were removed, to achieve extraction rates of up to 90%. The zone of working retreated back towards the shaft, leaving collapsed ground, known as goaf, in its wake.

Remains of buildings at the old Skelton Park mine.

Removing the pillars is what made mining so dangerous, because the result was rapid collapse of the roof. The miners used timber props as temporary, sacrificial, roof supports. In each working sector a weak or weakened timber prop was installed so that it would break at the first sign of roof movement, at which point the miners would rapidly depart, leaving the entire area to collapse. This was a hazardous and often fatal process in the days before movable steel roof supports became available.

The collapsed roof in retreat mining inevitably created ground subsidence. So some areas of the mines were left with pillars intact, in order to avoid ground movements that would damage the access shafts, or to prevent surface subsidence of houses and infrastructure. Farmland was allowed to slowly subside over the goaf with little ill effect.

The Kilton ironstone mine closed in 1963, and its double cone of tipped waste shale is now known as Kilton Hill.

Mines at Eston, Skelton and Loftus

Eston was the first, the largest, the most famous, and arguably the best of the Cleveland mines. After starting with small quarries along the outcrop in 1850, underground working commenced in 1852, and continued until 1949. The ironstone was extracted through numerous drifts along more than five kilometres of outcrop, which lay midway up the steep scarp face rising to Eston Moor. At times it had 700 miners who were taking five metres of payable ironstone; the Main Seam was four metres thick, with only a thin shale bed separating it from the underlying Pecten Seam. By the time it was closed, Eston Mine had yielded 63 million tonnes of iron ore.

At Skelton, east of Eston, a group of mines had North Skelton as the key player; it produced 25 million tonnes of ore during its lifetime of 92 years, and was the last of Cleveland's iron mines to close, in January 1964. It worked the Main Seam, which was around three metres thick, but this lay in the trough of a syncline, so its shaft was the deepest in Cleveland, reaching 230 metres deep. Underground roadways

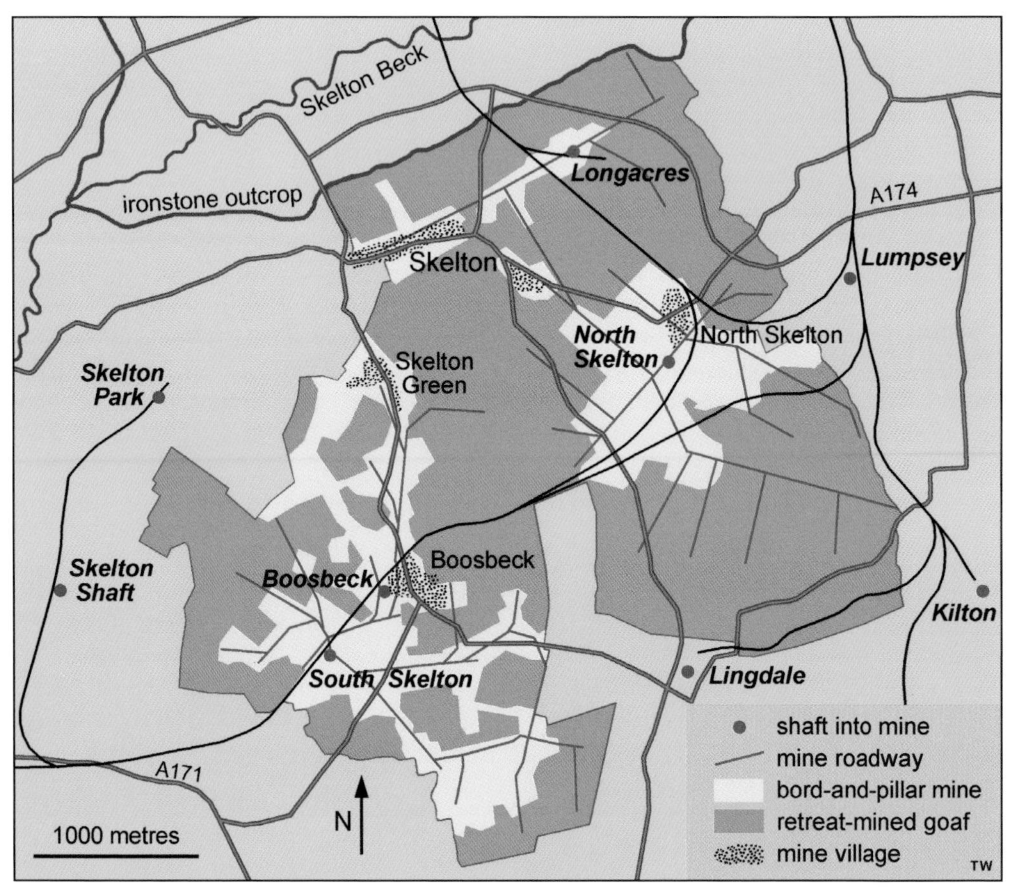

Ironstone mines beneath Skelton. The extent of the mine workings is taken from an undated map compiled between 1925 and 1947, and it is likely that the goafed areas were extended further before mining ceased in 1964; pillars would have been left in place to ensure stability beneath the villages and around the shafts. The villages are shown as they were during the period of mining; roads are as they are today. The railway tracks have since been removed, except for the northern loop, which remains in use as the mineral line to the Boulby potash mine. Workings of the adjacent mines were extensive, but are not shown on this map. The ironstone outcrop is that of the Main Seam of the Cleveland Ironstone.

A metre of mudstone was taken from the floor to create a working height in the main roadways of an old mine in the Avicula Seam of the Cleveland Ironstone in lower Eskdale; the exposed ironstone has a weathered and leached crust of grey clay minerals.

aligned down the dip of nearly 4 degrees were inclined steeply enough to run rope haulage by gravity, with loaded tubs descending towards the shaft bottom while pulling empty tubs back up to the production faces. Galleries also connected to adjacent mines, none of which reached as deep as North Skelton, but those to the south found the ironstone losing thickness and also gaining a shale middle.

Before it became part of South Skelton, the Boosbeck mine worked two metres of ironstone below a shaft that was 90 metres deep. This mine was notable for creating Cleveland's worst case of subsidence damage, when in 1883 some 200 of the village's houses had to be demolished after serious ground movements. Pillars of ironstone had been left in the parts of the mine directly below the village, but had been extracted to leave collapsed goaf almost immediately to the south. It is now well documented that a zone of collapsing ground flares outwards from an area of roof-fall within a mine, so that a larger area of the surface is affected by ground subsidence. This should have been known then, and subsidence of the Boosbeck houses was deemed in court as being due to 'improper mining'. Subsequently, the mine closed after an inrush of water in 1887, though some remaining ironstone was worked from the South Skelton mine, which remained dry as it lay further up the dip.

East of Skelton, the Kilton mine also worked the Main Seam; it was the last of Cleveland's mines to use horses in its underground transport, until 1951, when they were replaced by diesel locomotives. The main tip heap at Kilton is now known as Kilton Hill. It is protected as an item of industrial heritage, so, unlike the majority of mine tips, will not be removed or flattened; it now stands as a fine example of a landform with its geological origin and processes rooted in the geological agent known as mankind.

Third of the giant ironstone mines was Loftus. Back in 1847 drift mining started in the three-metre-thick Main Seam along both sides of the narrow Skinningrove valley that leads down to a small port. At first the ore was shipped out by sea, but in 1874 two blast furnaces were built as the core of an iron works on the hill west of the valley. This site later adapted to steel production, but is now only a rolling mill. Meanwhile the mine extended east and south beneath Loftus, yielding a huge tonnage of ironstone, before underground operations ceased in 1958. By then the mine output included shale that came from a middle thickening towards the south. The shale waste was separated on the surface and then tipped into the tributary valley of Deepdale; eventually this was completely filled with the waste stone, thereby becoming another landform, or a landform lost, due to the geological agent mankind.

The site of a short-lived ironstone mine destroyed in 1858 by renewed movement of the Wrack Hills undercliff on which it stood west of Runswick Bay.

Mines of the outer fringes

Beyond the core of ironstone mining between Eston and Loftus, the Cleveland ironstone belt extended to various outlying sites with their own distinctive mines. Perhaps the most spectacular were in Rosedale (*see* page 116), but the Whitby coast and Eskdale both have histories of successful iron mining.

Working a metre-thick seam of Cleveland Ironstone (leached to grey at its surface) was only practicable when some barren roof rock was also taken out.

The earliest ironstone mining on an industrial scale (still of modest size, but greater than tiny diggings for local forges) was along the coast both east and west of Staithes. There it was mainly the thin Dogger ironstone that was exposed in the cliffs, where it was hacked out and loaded into boats that took it to Tyneside. A scatter of workings even extended to drifts following the seams underground. Port Mulgrave shipped ore from the Dogger before developing a little further with shafts on the quayside down to the metre-thick Main Seam around twenty metres below sea level, which was worked westwards towards Staithes.

North of Runswick Bay, the undercliff known as Wrack Hills was in 1856 the site of a mining operation with drifts into the Dogger seam near the cliff top, and shafts to the Main Seam that lies below sea level. Kilns, furnaces and a small jetty were built, but mining had barely started when the whole site was virtually destroyed by a landslide in 1858. Such is the way with undercliffs, which are essentially old landslides prone to renewed ground movement.

Easternmost of the ironstone mines was the Whitehall Pit, in the southern suburbs of Whitby. In 1859, a shaft was sunk twenty metres deep to reach the Dogger seam, but it found difficult ground that was also prone to flooding; machinery from the failed mine was sold off just two years later.

In Eskdale, the Main Seam of the Cleveland ironstone rises to outcrop on the crest of the Cleveland Anticline, but is too lean, too thin and too much split by shale middles to be worth mining. However, the underlying seams, the Pecten and the Avicula, named after their fossils, each offer around a metre of good ore. Most of the mines were small, with the Hollins and Hay's mines in the south side of Eskdale as the largest. For 30 years these supplied the ironworks built in 1862 at Grosmont.

The Eskdale mines were largely worked by drifts in from outcrop, but their seams are overlain by shale that created unstable roofs; some wider bords were stabilised with stacked waste stone, but there was no scope for taking out the pillars during retreat. Mines in the valley were started in 1835, but all had closed down by 1877, though the Hay's mine was re-opened in 1906 and then worked for just nine years. An outlying mine, on the eastern side of Glaisdale village, was unique within the valley by having a shaft 75 metres deep to separate workings in the two seams that are locally ten metres apart. The mine fed its own iron furnaces just across the River Esk.

With all the ironstone mines of Cleveland closed down, most of their sites are now long forgotten and have slowly blended back into the landscape. None of the ironstone quarries was large, and the many small quarries now appear little different from scars or crags along the outcrop of a slightly harder rock. At some sites the ramps of tipped waste stone, or the waste slag from furnaces, can be recognised even where they are all grassed over, though many tips were removed because they were

A small roundabout has been created by the concrete cap on the 80-metre-deep shaft that once gave access to the ironstone mine at Aylesdalegate.

easy sources of hard material for construction projects. Shafts have been capped, and most drifts are either collapsed or blocked off. Mine buildings survive at some sites, notably at South Skelton and Skelton Park, and the magnificent remains of the three sets of kilns in Rosedale are now preserved and protected. The greatest relics of mining are the railways, now with all the rails gone, but many of them providing ready-made footpaths and cycle tracks.

One mine that might have made a larger impact on the landscape was Roseberry. This extracted more than two metres of ironstone from the Main and Two Foot seams beneath the splendid isolated hill known as Roseberry Topping. Mining dates back to 1871, but lasted only a few years, before a second phase ran between 1906 and 1926. Within that time, in 1912, a large landslide scarred the southern face of the hill and left the steep wall on the sandstone crag at the summit. The landslide was blamed on the mining at the time, but the more significant cause was probably the heavy rainfall prior to the failure, with the mining being only a contributory factor by its prior disturbance, and weakening, of the ground. Little remains to be seen of the Roseberry mine, but the jagged profile of Roseberry Topping is still a conspicuous feature in the landscape, and could just about count as a relic of the bygone mining industry.

Ironstones of Rosedale

Close to dead centre of the North York Moors National Park, Rosedale is a delightful valley, with a floor of farmland overlooked and almost completely surrounded by heather moors. It is a site of rural peace today, but it has a dramatic history of industrial activity that left its mark on the landscapes. From an outcrop high in the slopes all around the northern part of Rosedale, a bed of ironstone has yielded more than ten million tonnes of valuable iron ore.

The little village of Rosedale Abbey is distinguished by never having had an abbey. Long ago there was a small Cistercian priory, but that was demolished before its

ABOVE Though now grassed over, the excavated face, the working floor and the small tip heaps combine to define the old quarry in Dogger ironstone set into the western side of Rosedale's tributary North Dale.

LEFT The ironstone mines of Rosedale. Surface mining was extensive along the Dogger outcrop, but only the main sites are marked, and they probably all had small drift mines into the hillside behind them, though any records of these are minimal. Track-beds of the old railways have been graded to form the new cycle trails and footpaths.

OPPOSITE Preserved remains of the old ironstone kilns at Scar Top, standing high on the western side of Rosedale; the uneven slopes in the foreground, now covered with grass and heather, are on the old banks of waste and burned stone.

site was occupied by the village church in the 1800s. There are records of forges in the valley, producing iron tools for the local farmers as early as the 1200s, and it is likely that these sourced their raw material from a scatter of small quarries into the dale's ironstone outcrop. Local resources were the most useful in olden days of minimal transport, and coal was extracted from thin seams that could be found beneath many parts of the surrounding upland. There were even a few small mines working the precious jet to supply the artisans of Whitby, though none lasted until 1900. It was all small-scale mining, until things changed after 1851.

The remarkable magnetic mines

South of Rosedale village, a little quarry above Hollins Farm was yielding hard stone that was used for improving the state of the local roads. Then in 1851, William Thompson travelled in from Staithes looking for sources of jet, and it was he who recognised that some of the local roadstone was magnetic, and traced it back to the quarry. The rock proved to have an iron content of nearly 50%, which was better than the ironstone then being mined in the Cleveland Hills. In 1853, Rosedale ironstone mining started in earnest. The quarry was expanded, drifts were worked into the hillside,

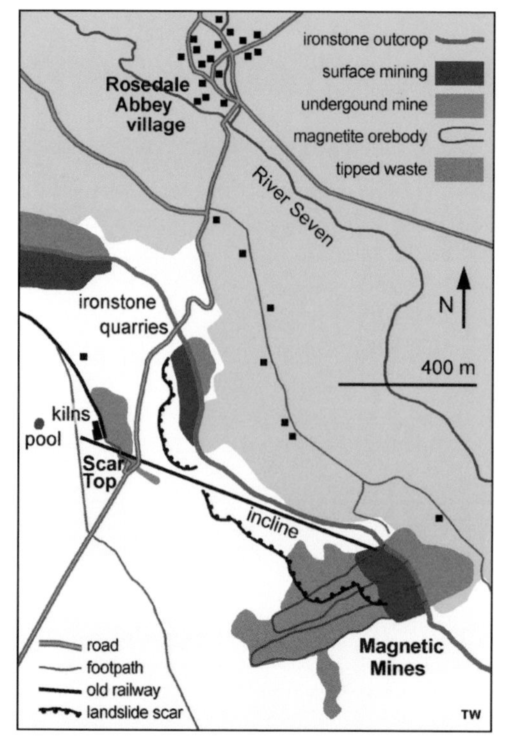

Western Rosedale. Landslide scars mark the upper limits of slopes that have slumped into the old quarries on each side of the road up to Scar Top. Underground workings are shown at the Magnetic Mines only. Farmland is green, moorland is yellow.

production increased, and men arrived for the work opportunities in what became known as the Magnetic Mines.

The rural population of Rosedale had been a few hundred, much as it is today, but it climbed to more than 3000 during the 1870s. It is often claimed that this was known as the Yorkshire Klondike, though that epithet must have come later, as nobody had heard of Canada's Klondike prior to the madness of its gold rush in 1897.

Sending the iron ore by horse and cart to the railway at Pickering had its limitations. In those days before motorised road transport, railways were the only means of bulk transport. So by 1861 a railway had been built, in the space of just fourteen months, from an existing line at Battersby, over the Cleveland Hills and along the upper slopes of Rosedale to Scar Top, which was up the hill from the mine. Subsequent additions took the track length to more than 30 km, all just for the iron ore. The railway surmounted the scarp face of the Cleveland Hills by means of the cable-hauled Ingleby Incline, 1500 metres long and rising 220 metres. Descending trains loaded with ore pulled up lines of empty wagons, with braking control on the winding drum at the summit.

Another incline, 900 metres long, had to be steam-powered to haul narrow-gauge ore trucks 100 metres up from the mine to Scar Top. There the ore was roasted in the giant kilns that still stand as great landmarks. Also known as calcining, this roasting removed the carbonates and volatiles, and thereby reduced the weight of the ore, which was critical in reducing tonnage costs on the railway to the iron foundries of Teesside. Coal to feed the kilns first came from the many small mines in the thin coal seam

The bench along the hillside that marks the line of old ironstone workings below Scar Top on the west side of Rosedale.

along Blakey Top, but was also later brought in from Teesside in some of the railway wagons by then emptied of ironstone.

The magnetic ore of Rosedale is something of a geological freak. It occurs in just two boat-shaped orebodies, each around 400 metres long and about 60 metres wide, that lie so close to each other that the initial two mines became connected into one. The magnetic ore reached nearly twenty metres thick along the centres, tapering to nothing on each side. It occupied channels that were cut into the underlying mudstone, and were then capped by the bed of leaner Dogger ironstone. The ore is virtually a black or rusted conglomerate, a jumbled mass of sandstone blocks mixed with ooliths of magnetite (iron oxide) and chamosite (iron silicate), set in a matrix of siderite (iron carbonate). Its iron content of 40–50% is greater than in any other sedimentary ironstone in the Jurassic rocks of Britain. With the mines now inaccessible, and with no surface exposure, debate continues over the origins of this ironstone (*see* page 13).

The two orebodies were completely worked out by 1879, and no other comparable channel of magnetic ore has been found. However, by 1879 the mining industry of Rosedale was already well into its second phase.

With the railway in place, effectively paid for by the riches of the magnetic mines, there was renewed scope for exploiting the lower-grade ironstone that is much more extensive within the Dogger beds of Rosedale.

The stone kilns outside the East Mine in Rosedale's Dogger ironstone.

Leaner iron ores in the Dogger

This ore is similar to the Cleveland ironstones, with chamosite and siderite providing most of the iron content, which reaches little over 30%. It varies in thickness and quality, from an unworkable metre of lean stone, to a profitable five metres of good ironstone. There were many small quarries and short drift mines along its outcrop, on both sides of Rosedale, into North Dale, and over in the adjacent Farndale; but only two areas of the Dogger ironstone could support larger scales of profitable mining.

The mine known as Sherrifs Pit was intentionally sited next to the railway, with a shaft sunk 80 metres deep into a thick zone of ironstone reaching in from the outcrop. It was worked from 1857 until 1911. Rather larger was the Rosedale East Mine, which warranted a branch of the railway round to it in 1865, along with two sets of calcining kilns. With shafts only required for ventilation, this was worked entirely by sub-horizontal galleries that emerged at the outcrop just above the kilns into which the ore was tipped. But even its great bed of ironstone eventually thinned to sub-economic levels, so the mine and its kilns ceased operations in 1926.

With its ironstone all worked out, the last train departed from Rosedale in 1929, and the valley slowly returned to its rural roots. Rail lines were recovered for scrap, and the track-beds were subsequently given gravel surfaces for easy walking and cycling. The last of the railway cottages at Blakey Junction were demolished in 1955. The notoriously steep road due south from Rosedale Abbey is still known as Chimney Bank; it took its name from the 30-metre-tall chimney that stood beside the winding house at the top of the railway incline, until it was felled in 1972. Only the adjacent reservoir remains on the empty moor.

Rosedale is now a quieter place. There is almost no trace of the mines, and none that is accessible – just a few scars that can be recognised as old quarries. But the skeletal ruins of the old kilns remain, and are preserved as grand reminders of times gone by.

Deep mines in the evaporite beds

Only seen at outcrop beyond the margins of the North York Moors, the Permian rocks that extend deep beneath the Moors have proved to be important resources of valuable minerals. The outcrop of the Upper Permian beds traces the western side of the Vale of York and then curls round to meet the coast north of Middlesbrough. It is something of a coincidence that it follows so close to the shores of the Permian sea in which the sediments were deposited. This was an inland sea on the contemporary super-continent, standing below sea level within a hot desert not far north of the Equator. Known as the Zechstein Sea, it extended from England across to Poland. It had inflows of river water from its hinterland, but, more importantly, sea water periodically poured into it from the open ocean, through a descending valley that lay between Norway and Greenland. Being a closed sea, its waters were only lost by evaporation. Salt and other evaporite minerals were thereby precipitated, and accumulated to great thicknesses over a period of about five million years. These have been mined extensively in northern Germany; 'Zechstein' translates from German as 'mine stone'. However, the valuable minerals were not so easily found in Yorkshire.

When natural waters evaporate, minerals are deposited in a sequence that is determined by how much concentration, due to water loss, each needs to reach saturation and be precipitated. The normal sequence starts with limestone, followed by gypsum, then halite (rock salt), and finally a mixture of potassium salts. This is repeated in five cycles that developed in the Zechstein beds. As more water was lost within each cycle, the level of the inland sea declined, and its area shrank. Ever-changing levels and extents of that Permian sea dictated the distribution of the valued minerals beneath the North York Moors.

The area of the Zechstein Sea extended to the edge of the Pennine uplift, so the Permian Magnesian Limestone has a long outcrop on the western edge of the Vale of York. With nearly the same extent, gypsum has caused subsidence problems at Ripon, and also relates to the anhydrite at Billingham (*see* below). Rock salt did not reach as far in a shrinking sea, so is not seen at outcrop, though it occurs beneath Middlesbrough. Then the potassium minerals were deposited in a shrinking sea that reached westwards only a little beyond Whitby, where they now lie at great depth due to the dip of the rock sequence towards the North Sea basin.

Beds of salt were discovered beneath Middlesbrough in 1862 when they were intersected by a borehole drilled in search of fresh water. A number of small brining operations then started up, pumping water underground and bringing brine to the surface. But salt later took second place to the anhydrite found in the same area.

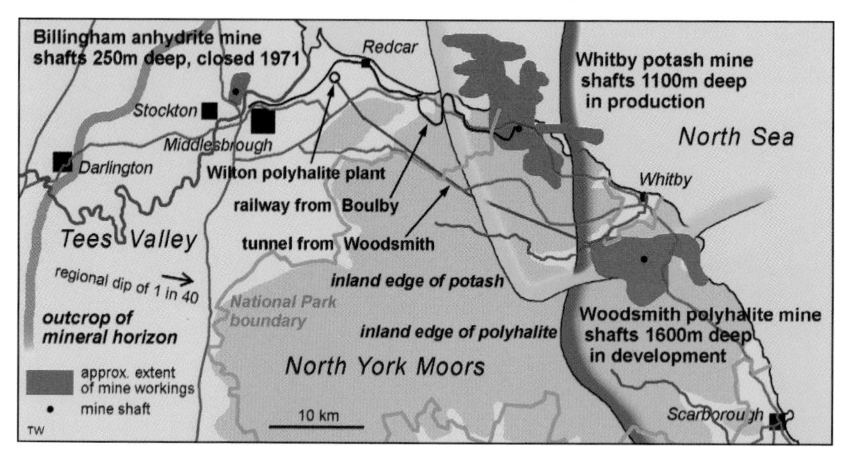

Features of the three deep mines that extract the evaporite ores beneath and near the North York Moors.

Anhydrite at Billingham

In 1917, a massive chemical plant was built at Billingham, across the river from Middlesbrough, primarily to produce ammonia. Then, some years later, boreholes encountered anhydrite directly beneath the site. Calcium sulphate had been deposited in the Zechstein Sea as gypsum (the hydrated form, $CaSO_4.2H_2O$). However, when gypsum is buried to depths of about 500 metres, the water is driven off to form anhydrite (simply $CaSO_4$); then when erosion removes its overburden, anhydrite rehydrates back to gypsum at depths around 100 metres. At Billingham, the mineral is 250 metres down, so still exists as anhydrite, and this was hugely valued to produce sulphuric acid and ammonium sulphate fertiliser at the chemical works directly above.

From 1927 to 1971 the mine took an annual million tonnes of anhydrite from a gently dipping seam nearly eight metres thick. Large machines and trucks could operate in roadways six metres wide, worked in two directions to leave square pillars thirteen metres wide. This yielded 51% extraction, and the pillars of intact mineral, which is stronger than concrete, ensured there was never any ground subsidence. Long after it was abandoned, the mine is still totally stable, and it was deemed suitable for storage of radioactive waste until a public outcry reminded planners that it was beneath a major centre of population. Despite its potential to accept a lot of non-hazardous waste, the mine has since remained unused, and its valuable empty space went up for sale in 2022.

Potash at Boulby

Further east from Billingham, the dipping Zechstein evaporites lie at greater depth, but have valuable potash beds lying just above the thick rock salt. Several metres of potash mineral were found by accident in 1939 in a wildcat oil-exploration borehole 1500 metres deep at Aislaby, near Whitby. The eventual outcome of that discovery was a pair of shafts 1000 metres deep to develop the mine at Boulby. This went into production in 1973, and now extracts three million tonnes of mineral each year.

Most of the potassium occurs as the mineral sylvite (KCl, potassium chloride), but this is mixed with halite (NaCl) and various other minerals. The sylvite forms about one third of the mined ore that is known as sylvinite in a bed about seven metres thick. From the foot of the shafts, roadways are driven in a bed of stable halite some eight metres below the bed of weak sylvinite. Inclines rise to working panels in the sylvinite where the mineral is excavated by continuous miners. These are mobile machines each with a rotating toothed cylinder mounted on a boom; they just grind their way through the ore, with debris falling on to an integrated conveyor belt. They extract a thickness of nearly four metres of mineral from headings between pillars that are left in place. However, beneath more than a kilometre of overburden the sylvinite deforms, so the headings experience floor lift and roof collapse that renders them inaccessible after about six months, when they are sealed off after production has moved to a new panel.

The coastal lowland behind Staithes, with the buildings of the Boulby potash mine at the foot of the hills.

A hand specimen of the ore mined at Boulby, a mix of sylvite and halite, coloured red by iron oxide.

Boulby's bed of sylvinite varies in thickness, and the mine is currently running low on proven reserves. So it now has an incline down to a lower bed of polyhalite that is extracted while there is market demand before the nearby Woodsmith mine goes into full production.

The mine now extends for 11 km from the shafts, mainly out beneath the North Sea, reaching depths of 1400 metres, far beneath the sea bed. Miners travel to their working sites in minibuses that are lowered intact down the shaft and then trundle round the spacious roadways. Boulby is a spectacular operation, totally unlike the coal mines of earlier years.

A mine within the National Park was always going to be controversial, but it was deemed that Boulby's available potash deposits were of national importance, so could overcome planning restrictions within the Park. A large surface structure was acceptable; it is inevitably conspicuous, but is neater, tidier and more compact than at most mine sites. The mineral processing plant and storage sheds are large, and the shaft headgear is inside a tall structure, but there are no tip heaps. The low-value halite produced by making the underground roadways is exported to be used as road salt, and any other waste is returned underground and tipped in abandoned panels. All the mineral goes out by rail, on the tracks restored to the active line that serves Saltburn-on-the-Sea, so there is no unwelcome road freight. The mine at Boulby may not be ideal in a National Park, but it is a major asset and employs a thousand people, all dictated by the geology.

Polyhalite at Woodsmith

Exceeding the Boulby potash mine in so many ways, the Woodsmith Mine has an even more valuable resource, lying at even greater depth, and will make even less impact on the landscape of the National Park. It is planned to go into full operation in 2024, and should ultimately produce an annual ten million tonnes of mineral during a projected lifetime of 100 years: it will be a world-class mine, with major value in its exports.

Most of the mineral taken from Woodsmith will be polyhalite, a complex hydrated sulphate of potassium, magnesium and calcium. This is a valuable component for fertiliser, as it contains multiple nutrient sulphates and is also free of the undesirable chlorine. The bed to be worked is 70 metres thick, so proven resources are immense, and there may be more to develop out beneath the North Sea. It lies within an early cycle of the Zechstein evaporites, beneath that with the Boulby potash and the Billingham evaporite. Along with the gentle dip to the southeast, this accounts for the mine's great depth: after those in South Africa and Canada, it will be the deepest mine in the world.

Fifty years on from Boulby, environmental restrictions within the National Park have become more stringent for the Woodsmith development. Located five kilometres south of

A continuous miner, with its boom-mounted, toothed drum, of the type that will grind out the full thickness of the ore bed in the Woodside polyhalite mine

Whitby, it is surrounded by woodland, is not overlooked by hills, has no tall structures, and has no tip heaps; most visitors to the Park will not even know it is there. At the site, two shafts descend nearly 1600 metres from headgear and winding houses that are largely below ground level. Only miners and machinery will enter through them. Mineral will be carved out using continuous miners (as at Boulby), which produce broken rock direct on to a conveyor belt. From the live workings it will be hoisted up to a depth of 340 metres, where it will be tipped into silos carved out of the bedrock. From those it will be fed on to conveyor belts running through a tunnel 37 km long.

This tunnel is the key factor in the mine's low environmental impact. It emerges at Wilton in the industrial zone of Teesside, where a mineral processing plant can feed product direct to ships in the adjacent port. Opportune geology means that the gently graded, six-metre-diameter tunnel is excavated almost entirely within the gently dipping Redcar Mudstone, which is close to ideal for penetration by a conventional tunnel-boring machine. The mine at Woodsmith is a truly exciting project that will become a national asset, with negligible impact on the National Park.

Coal and gas

All across the Moors, seams of Jurassic coal are common enough among the deltaic rocks within the Ravenscar sequence, but nowhere matches those of the Carboniferous in western Yorkshire. They are nearly all less than half a metre thick, and the coal is of poor quality. It was mainly dug for burning in nearby lime kilns, though some was used on home fires. Numerous small mines were worked as bell pits, mainly during the 1700s and 1800s, though some were briefly re-opened during the hard times of the General Strike in 1926.

A bell pit was a hand-dug shaft about a metre in diameter and rarely more than ten metres deep. Once the seam was met, the coal was dug out in all directions, and small failures of a weak roof rock flared the cavity out to give it the shape after which it is named. Galleries might extend a few metres in any direction, but with underground transport limited to a sled dragged over a muddy floor, it soon became easier to dig another shaft not far away. Many bell pits are still recognisable by their ring of debris now covered by heather, but it is best not to descend into their central hollows. Survival of the debris ring means that the abandoned shaft was not backfilled; many were merely covered by any bits of spare woodwork, which might now be centuries old: say no more.

The choked shaft of an old bell-pit mine into a thin coal seam on Blakey Ridge, above Rosedale.

Most of the workable coal lies within the topmost beds of the Cloughton Formation, so bell pits were sunk through the overlying sandstones and mudstones of the Scarborough Formation where erosion had left only a thin cover on the moorland plateaus. There are probably more than two thousand abandoned coal pits scattered across the high moors. Seams varied in thickness, so where a seam was found to be locally thicker a cluster of bell pits would follow to gain from the richer pickings. There are the remains of dozens of bell pits in groups on the high moors around the head of Rosedale (*see* map on page 116), where coal was in demand to feed the ironstone kilns until it was cheaper to bring in supplies from the Durham coalfield. Very much in the minority, a few small drift mines worked locally thicker coal seams in from outcrop; notable were those in the Clitherbeck valley, tributary to Eskdale near Danby, where some larger tip heaps are still recognisable to indicate the sites of adjacent drifts and shafts now sealed off.

The low mound on the skyline is a ring of debris round one of many old bell pits that were sunk along Blakey Ridge to extract coal from beneath the heather moor.

Successors to coal are oil and gas, and eastern Yorkshire has resources on a modest scale, though almost exclusively of gas. Methane is the main constituent of this natural gas that has its source at depth in the Carboniferous Bowland Shale and Coal Measures. From there it has migrated up into Permian sandstone and limestone reservoirs, where it has been held in place by the evaporite beds that form an effective caprock. Payable gas fields occur in buried gentle anticlines of the reservoir rocks, beneath the Vale of Pickering and nearby areas. Gas was first found in a well that was drilled near Aislaby, in Eskdale, in 1937. Since then more than a hundred wells have been drilled; most have proved to be dry, but eight have gone into production.

The Eskdale well fed gas into the town supply in Whitby, but lasted only from 1960 to 1967. Soon afterwards, a well at Lockton supplied gas to Scarborough for three years, but the region's largest producer has been at Kirby Misperton, in the Vale of Pickering. Since 1995, this has supplied gas to the Knapton power station in the Vale of York; however, it will cease production during the 2020s. There remains potential to extract further gas from this well, but only after yields can be enhanced by hydraulic fracturing of the reservoir rock, in order to increase its permeability and gas flow. This would be standard procedure in any oil or gas well, but since becoming known as fracking it has attracted opposition, and any further development is on hold. As an alternative there may be significant potential for using the 3000-metre-deep borehole for geothermal energy, as ground temperatures reach 90°C at the bottom of the hole. The same might well apply to three other wells in the Vale of Pickering, all of which produced gas during the 1990s but are now closed down. Another gas field that is now depleted lies beneath Caythorpe, just west of Bridlington.

The most productive reservoir rock has proved to be the Kirkham Abbey Formation, a 70-metre-thick dolomitic limestone once known as the Upper Magnesian Limestone; it

Small tip heaps of broken waste shale are relics of the many small coal mines that were worked around the Clitherbeck Valley, tributary to upper Eskdale at Danby.

lies beneath the beds of salt and potash deep within the Permian sequence. Exploration for gas has continued to target this bed, which was a near-shore sediment, so is restricted to a narrow belt extending deep beneath the Moors and Wolds. A major recent discovery was made at a depth of 2000 metres beneath West Newton in the southern part of Holderness. Tests on the two exploration wells indicate that this could be the largest on-shore gas field in Britain; it is scheduled to go into production in 2026.

Oil is overshadowed by gas in eastern Yorkshire. A deep borehole at Fordon, near the eastern end of the Wolds, found oil in Carboniferous sandstone, but the tiny amounts offer no prospect of commercial extraction. Small volumes of oil are likely to be a by-product from the West Newton gas wells, but there is still little prospect of more oil coming out of eastern Yorkshire.

Out beneath the North Sea, the Easington gas fields lie 70 km east of Holderness, at the western end of the great swathe of gas fields extending from the Gröningen site that has been an important producer in Holland since the late 1960s. Easington's fields were once important producers, but all those close to the Yorkshire coast are now depleted. On the adjacent Holderness coast, the Easington terminal receives gas from many parts of the North Sea and uses some of the depleted fields for gas storage, by pumping gas back down the original wells and into the available reservoir rocks. The terminal has additional gas storage in large caverns 1800 metres beneath ground level at both Aldbrough and Atwick, either side of Hornsea. These were purpose made by drilling down into the thick bed of Permian salt and creating solution caverns by conventional brining – pumping water underground and extracting the brine until the caverns reach the optimum size that can remain stable.

The next stage in the story of Yorkshire gas may or may not involve extracting shale gas from the Bowland Shale. This thick unit of Carboniferous shale offers a major resource some 2000 metres beneath the Vale of Pickering. However, it is the type of gas that is held tight within the rock and can only be released after the rock has been fractured by fluid injection at high pressure. This is the process colloquially known as fracking, which remains controversial because of the on-going disputes over the scale of its side effects, notably aquifer contamination and earthquake generation – which is why it may or may not happen in the Vale.

Jet ornaments

Jet is fossilised wood. It is essentially a type of coal, but is a special variety that was formed in a specific environment, away from the usual coal-forming swamps. Jet's distinctive properties are that it is absolutely pure black, it is amorphous and tough enough to be carved with great intricacy, and it takes a high polish. It therefore makes an unusual gemstone, though it is not hard, so requires some care in handling; it has a long history of being carved into small ornaments.

Jet started life in Jurassic times as branches and trunks of trees that fell into rivers and were carried out to sea where they became waterlogged and therefore sank, to be buried in the seabed mud. In the mud's anaerobic, reducing environment, the wood lost its volatiles and retained its carbon (in contrast to wood that decays in air and loses its carbon by oxidising into gases). Now compressed and fossilised, the fragments of trees were on their way to becoming lumps of coal, except for a fortuitous situation within those Jurassic sediments. An unusual set of chemical processes involving hydrocarbon fluids and bacterial action, gave the material the texture and structure that is specific to jet. The same wood when buried in freshwater sediments of the delta transformed into a slightly different material, known as 'soft jet', which cannot be worked as well as the 'hard jet' from marine sediments. Then, by good fortune, the fossil wood was not subsequently invaded by siliceous fluids to create the petrified forests that are known elsewhere.

By way of its origins the jet does not form beds, but instead occurs only as scattered lumps within the ten metres or so of mudstones that constitute the Jet Rock. Most lumps are not large; anything more than about ten centimetres across has rarity value. The best material occurs as flat slabs, known locally as planks, which were the strong bark of tree trunks with opposite sides flattened together after all or most of a pith core had rotted away. A tiny minority of jet fragments retain the pattern of the original bark. Some of the soft pith cores that are characteristic of the Araucarian trees were replaced by mud prior to complete burial, so they created lumps that are known as 'cored jet'.

Though it has not been the main source of jet, Whitby is famed as the place where jet jewellery and ornaments have been made in a host of workshops that have been there ever since the heydays of jet in the late 1800s. Jet can be found on the beaches around Whitby, but this is material eroded from the outcrops of Jet Rock in parts of the coastal cliffs between Saltburn-by-the-Sea and Robin Hood's Bay. Numerous small drift mines were dug into those same cliffs, mainly to the west of Sandsend, but most of the jet mines lay far inland.

Fragments of jet found on the tip heap of an old mine.

A museum specimen of jet that retains some of the structure of its orignal tree bark.

The largest numbers of nineteenth-century jet mines lay along the Jet Rock outcrop high in the scarp face of the Cleveland Hills between Guisborough and Swainby, and also just over the crest of the Hills around the upper slopes of Bilsdale, Raisdale and Scugdale. Jet mines were never large operations, and all were worked as drifts in from outcrop. A typical mine had an extensive grid of narrow passages, nowhere reaching more than 100 metres in from the outcrop. They were cut with hand tools, and the mudstone was too hard to work beyond the zone of near-surface weathering. Once the passage was cut, it was easy to bring down the roof; lumps of jet could be picked out, then more roof was brought down by miners standing on the pile of debris. This could be repeated to work up through the richest of the Jet Rock in its top few metres, until the mine was abandoned with a stable roof of Top Jet Dogger, a thin bed of harder ironstone (which was not worth working for its iron).

A tall and narrow gallery typical of the small drift mines working for jet in the hills above Guisborough.

Around the head of Bilsdale, just below the Wainstones, the jet was so abundant that the miners were reluctant to leave any in pillars, so they worked quarries back into the entire outcrop until there was too much overburden to remove. Though degraded and overgrown, old tip heaps of barren shale can still be recognised to indicate the extent of bygone working. Bilsdale had 42 miners extracting its jet during 1871. Picked out of the excavated

Medallions carved in jet: King Oswy and his Queen Eanflaed on the left, and Prime Minister Gladstone on the right.

The Jet Miners inn still stands in Great Broughton, though it is no longer the meeting place for jet miners from Bilsdale and jewellery craftsmen from Whitby.

Just below the rim of Raisdale, old workings for jet can be recognised by the overgrown tip heaps and the undercut scar that remains in the caprock.

mudstone, the good mineral was barrowed down to the Jet Miners Inn in Great Broughton, three kilometres away, where it was sold to buyers who had come over from Whitby.

Whitby jet is not unique to its namesake. The mineral occurs, and has been worked, in many parts of the world, though it could well be claimed that Yorkshire has provided gemstone jet of the best quality. By 1870, Whitby had 1500 of its residents employed in nearly 200 jet workshops, before the industry went into decline after cheaper, but softer, jet was brought in from Spain. Today's streets of Whitby are lined with shops selling the precious jet, but local supply is limited to what can be picked from the beaches. No mines remain active either nearby or in the Cleveland Hills. Irkutsk in Siberia is now a major source of raw jet, where it is known as gagate. Mines come and go, but trade and traditions continue.

Degraded workings of the Cleveland Dyke across the northern slopes of Goathland Moor.

The Cleveland Dyke

It is not strictly a mineral, because dolerite is a rock, but mines and quarries exploited the Cleveland Dyke in the past because it was a valuable resource of hard rock aggregate ideal for road surfacing. From the late 1700s onwards it was quarried to be trimmed into blocks, notably setts for paving roads, but after about 1900 most of the output was crushed into aggregate; this was still valued for surfacing roads because it is so durable, and it also binds with bitumen better than many other igneous rocks. The dyke extends from Fylingdales Moor to Great Ayton and beyond, though many sections are buried beneath alluvium or glacial till. Abandoned quarries scar almost the whole length of the dyke where it reaches outcrop, and the larger operations extended into underground workings.

Large galleries on two levels within the long-abandoned Sil Howe Whinstone Mine that worked the Cleveland Dyke beneath Gothland Moor.

What remains of the dyke and its workings are best seen on Goathland Moor where they are crossed by the road from Goathland to Sleights. The old quarry workings extend in a straight line westwards for a couple of kilometres along Whinstone Ridge. Never more than twenty metres deep, before they became unstable or flooded, these are now somewhat degraded and covered in heather and turf. There is little exposure of any dolerite, and the great linear ditch is a distinctive feature that sits well in the landscape and also recalls a geological and industrial heritage.

There is, however, more to the dyke that remains largely unseen beneath Goathland Moor. The old quarries were developed downwards into mine workings; these extend over a length of 1100 metres beneath an arch of rock left in place to support the floor of the surface cut. Active until 1950, this mine was named after Sil Howe, a round barrow of the Bronze Age that was only ever a metre high and is now unrecognisable where the moor

LEFT ***In the worked-out Cliff Rigg Quarry on the Cleveland Dyke, north of Great Ayton, heaps of waste stone have been moulded into a BMX bike track for the younger local residents.***

BELOW ***A young cyclist flies from a ramp on the BMX bike track created in the old Cliff Rigg Quarry.***

road crosses the line of the dyke's quarries. Underground, two levels of galleries, each up to fifteen metres high and wide, are slightly offset because the dyke is not quite vertical. Broken dolerite was taken down through the mine and out through an 800-metre-long, gently inclined tunnel that also served to drain the mine. From its exit, a tramway headed southwards, gently descending to reach the railway at Goathland for onward transport, and this can still be traced, winding its way across the featureless moor.

Further west and just off the edge of the Cleveland Hills north of Great Ayton, the Langbaurgh Ridge also traces the Cleveland Dyke. The old quarries, in a long straight line, are now largely filled with industrial waste and blanketed by soil, though they are still defined by their double line of bordering trees. This site was the last to take stone from the dyke, before closing down in the 1970s. It, too, had some large underground workings, but these are flooded and their entrances are buried beneath the landfill.

Now rather more visible are the old quarries eastward along the dyke, on the other side of the road to Guisborough. Cliff Rigg Quarry is a great gash in the hillside, though largely obscured by trees that have grown since the quarry's closure in 1918. A few small ribs of dolerite are still to be seen where good stone was left in place to prevent the weaker shales from slumping into the quarry. Cliff Rigg also extended underground with some large mine galleries towards the southeast, but these have since been sealed off. Stone for the local roads now has to come from distant sources.

From the abandoned dolerite quarries to the modern potash mines, such is the evolution of the extraction industries of the North York Moors. All have been dictated by the geology to become integral components of the varied landscapes that make up the moors, wolds, hills and vales of eastern Yorkshire.

CHAPTER 10

Villages of Stone

Yorkshire's Moors and Wolds share the honours of being home to a scatter of delightful villages that have long been established as integral parts of the countryside. Little of England's landscapes can now be claimed as entirely natural. Five thousand years back in time, the tree cover was almost ubiquitous, whether it was lush oak woodland in the valleys or sparse birch scrub on the high ground. Large-scale clearance of the trees started with Neolithic (New Stone Age) settlers, to yield both land for farming and also wood for building or burning. So wilderness has largely been replaced by agricultural terrains, where villages of stone buildings add much to the character of the landscape.

Surrounding the wide open spaces of the hill country, many parts of the lowland present far busier landscapes. These include the industrial towns old and new across the northern fringes of the North York Moors, and the string of resort towns spread along the Yorkshire coast; Whitby, Scarborough, Filey and Bridlington have been major seaside holiday towns since Victorian times. Each has its own distinctive features, but each is swamped by new developments that tend to be the same anywhere.

It is the villages and smaller market towns that have retained their character, each with its own individual style and history, and many bear the influences of local geology. For the earliest settlements a source of good water was essential, preferably from a reliable spring. Especially on the Wolds, where few surface streams can exist on the permeable chalk, many villages are on the spring lines. Dating from the tenth century, but now in ruins, Wharram Percy shelters in a valley in the north slope of the Wolds, just below where a spring pours forth from the base of the chalk, to feed the village pond that sits on the underlying clay. Large springs occur where chalk groundwater overflows on to the till cover of Holderness, and Driffield owes its Anglo-Saxon location to the water from its powerful Keld spring.

The village of Staithes crammed into the narrow valley that leads down to its harbour.

Whitby Harbour in the days when it was home to a thriving fishing fleet.

The ruins that remain of Whitby Abbey, built with Aislaby Sandstone in the early 1200s in a community that had already been established for 600 years.

In contrast, a lack of water saw some lowland villages established on areas of slightly higher ground: these were islands within the wetland, before drainage schemes increased the area of land available for farming. Kirby Misperton, on a mound of glacial till projecting through the lacustrine sediments that comprise the floor of the Vale of Pickering, and Aldbrough on a remnant of gravels atop the low-lying till of Holderness, are two of numerous early settlements guided by the region's geology.

Many coastal villages are sited where a natural harbour or a large beach could keep small fishing boats safe ashore between bouts of harvesting the sea. Settlements whose beginnings related to fishing generally developed into trading ports. Staithes is a lovely village crowded on to the flanks of a perfect natural harbour within its river mouth, though rather smaller than the mouth of the River Exe, which was key to the growing importance of Whitby. In contrast, the delightful village of Robin Hood's Bay (with no known connection to the buccaneer of Sherwood Forest) extends down to the water's edge with no harbour at all. When it was an important fishing centre prior to 1500, crews may well have hauled their boats on to a small beach above high-tide level, but this would have been lost to the influence of a slowly rising sea level, which has been relentless as a result of global warming ever since the Little Ice Age temperatures ameliorated soon after 1600.

Wharram Percy lies on the edge of the chalk Wolds, but its twelfth-century church of St Martin was built with calcareous grit from the Howardian Hills.

Houses in the lower village of Robin Hood's Bay are squeezed into the narrow valley of the King's Beck with the stream now in a culvert beneath the main street.

Mines can generate their own settlements, but these can fade away after mineral deposits are exhausted. Loftus and Skelton both originated at ironstone mines, and both have survived with new livelihoods, though terraces of old miners' houses can still be recognised. Not so in Rosedale, which was crowded with miners in the late 1800s, but now has only a small village with neither mine nor abbey. In complete contrast, the Moors' modern potash mines, Boulby and Woodsmith, have no need for pithead housing when miners can travel in from nearby towns. Different again was the thriving village known as Peak, above the southern end of Robin Hood's Bay, when it housed workers at the Peak Alum Works. After the demise of the alum industry in the 1860s, the village was rebranded as Ravenscar, with plans to develop it as a seaside resort to rival Scarborough, with convenient access via the new coastal railway. By 1902, streets had been laid out – but the project was doomed to fail, as visitors would have been faced with a long, steep hike to a rather paltry beach.

The origins of any settlement are commonly lost in history, especially where local industries, trades and culture have changed over time. But the fabric of the buildings, the churches and the houses tends to survive in the old centres of villages and towns. And this commonly relates to the local geology, which determined the sources of stone that inevitably replaced timber as the building material of choice. For many of the larger buildings, whether civic or private, best quality stone could be hauled in from afar, whereas most houses were built with locally available stone. This dictated the appearance of the villages and still defines their visual diversity, even where the bedrock is all but obscured by its mantle of soil and farmland. Each region therefore has its own character built into its villages and towns.

The church of St Peter in Hackness, built with a locally strong sandstone known as Hackness Rock, which is a thin part of the Osgodby Formation.

Stone of the North York Moors

Strong sandstone accounts for the presence of the upland moors, and the same rock provides a resource of good building stone. Towns and villages all across the Moors are built with the medium-grained, buff-coloured stone, either 'dressed' into roughly hewn blocks, or sawn to produce the blocks and slabs known as ashlar. With its early monastic origins, the remote and magnificent Lion Inn, high on the moors above Rosedale Head, is a fine example of an old structure built with blocks of sandstone that were sourced locally when transport capabilities were minimal in the early 1500s.

Pre-eminent among the natural building materials of the Moors is Aislaby (pronounced *Aizulbee*) Stone, a sandstone from the Saltwick Formation in the lower part of the Ravenscar Group. It is an excellent freestone, in that its homogeneous nature means that it can be cut or worked in any direction. Its main quarries are a kilometre west of the village of Aislaby, and a thousand years ago these produced the stone to build Whitby Abbey. Besides extensive local use in so many villages around Whitby and Eskdale, Aislaby Stone was also shipped south. In 1847 alone, nearly 25,000 tonnes of stone were recorded as despatched out of Whitby, after being hauled on ox-drawn carts for six kilometres from the quarries. It was used to build Covent Garden and to form the foundations for John Rennie's 1831 version of London Bridge (though the bridge's superstructure was built of Dartmoor granite, before its outer skin was sliced off and sent to Arizona to clad a concrete replica). Aislaby's early quarries were cut into the escarpment up from the Egton road, but these are now almost lost in woodland. Extraction has continued in the newer quarries reached from the main road to Middlesbrough. Their product is marketed as Eskdale Stone, and most of the output is as sawn blocks, which are now more in demand than the rough-hewn blocks of olden times.

Across the whole of the North York Moors, many beds of the Ravenscar sandstones were quarried in the past for local building works, although most of the best stone was taken from the lower beds (in the Saltwick Formation). Indeed, many quarries were in the same horizon as the Aislaby Stone, even where worked under local names. It was largely the better transport of the twentieth century that saw closure of most quarries, when production could be improved at larger sites where the stone quality was better. On the open moors, many of these little old quarries survive as rock scars that are indistinguishable from the natural landscape, whereas those in the valleys, notably all along Eskdale, are typically lost behind new growth of trees.

Extracting the valuable sandstone in a part of the present quarries at Aislaby.

ABOVE ***An old quarry on Great Ayton Moor that was abandoned when many of the blocks of its Saltwick sandstone had already been hewn into rough shape.***

RIGHT ***A substantial house on the outskirts of Whitby, built with blocks of various sandstones that were sourced locally from multiple beds in the Ravenscar Group.***

Softer beds were ignored, and parts of the Moor Grit were strong enough to yield aggregate to surface the early roads. Between those extremes, the sandstone was used extensively for houses and bridges, and it was only with the passing of time that a minority of the stone proved that it weathered rather badly in the exposed faces of buildings. Sandstone with a siliceous cement, such as the Aislaby, weathers well and retains a clean-cut face, but a calcareous cement could be lost to dissolution so that sand grains become detached and the stone surface is soon indented.

Some of the stone, notably in the Cloughton and Moor Grit units, splits into thin beds so that it is described as flagstone; this was worked for paving flags, but little of it was thin enough to be good for roofing, where a stronger and thinner material was required.

Above the Ravenscar sequence there are further sandstones that are generally softer, so are not widely used for building. A notable exception is the Hackness Rock, which was quarried at various locations near its eponymous village and was used to construct the Rotunda Museum in nearby Scarborough. One of Britain's oldest purpose-made museums, this was designed in 1829 by William Smith after he had famously produced the first geological map of much of England, and it still houses an important collection of Yorkshire's rocks and fossils.

Beggar's Bridge, built across the River Esk west of Egton in 1619, using blocks of stone sourced from a nearby quarry in one of the Jurassic sandstones.

Stone of the Cleveland Hills

Sandstones of the Ravenscar Group continue beyond the Moors to provide further resources of good building stone throughout the Cleveland Hills, with the best stone again in the lower part of the sequence. Houses and villages mimic those of the Moors, but one difference is in some of the old quarries, which had taller faces where they were cut into the steep northern slopes of the escarpment that forms the higher Cleveland Hills. Now abandoned and weathered, they blend in with the natural crags and have enough height to attract the more acrobatic rock climbers. Park Scar quarry above Kildale, and Beacon Scar above Swainby, are two sites where recreational climbers have taken over from quarrymen.

Guisborough Priory was built in the early 1300s, and is described as being made of blocks of Aislaby Stone. However, it is unlikely that stone was hauled nearly 30 km from the famous Aislaby quarry in those early times, when good sandstone, probably in the same bed, was available close by. A long-abandoned quarry up on Highcliff Nab is a strong contender as supplier of the stone. Whereas the Priory is now in ruins, the quarry face has become a fine rock crag, part of the landscape, and another venue for the climbing fraternity.

The ruins of Guisborough Priory, made of stone that was probably sourced from a quarry on Highcliff Nab, in the same bed as the Aislaby Sandstone but only three kilometres away, and all downhill to the Priory.

Typical lumpy terrains of grassed-over piles of waste stone inside a long-abandoned quarry near the crest of the Hambleton Hills above Kepwick.

Corallian stone in the Hills

Across the southern rim of the North York Moors sandstones and limestones of Corallian age form both the Tabular Hills and the Hambledon Hills. Villages and towns within this slice of countryside have building stone of top quality available at almost every turn, and the limestones yield a mellow stone that is more pleasing to the eye than the rather unexciting sandstones of the Moors to the north.

Both oolitic limestones (Hambleton and Malton), and also the intervening calcareous grits (Birdsall and Middle), have been used in construction, whether dressed into ashlar for houses, or left rough for walling. The Malton Oolite has yielded better and more widely used building stone; Thornton Dale and Helmsley are notable for their many fine buildings made with its oolites, though local sandstones also appear at both sites. In contrast, the Hambleton Oolite was largely burned to produce agricultural lime where it was extracted from the large Kepwick Quarry, high on the western escarpment of its eponymous hills.

Part of St Gregory's Minster, tucked away in Kirkdale, dates from around 1060, and was built with a mix of Malton Oolite and Coral Rag; both of these stones were probably sourced from the nearby quarry that subsequently achieved

ABOVE ***Helmsley has been built with a mixture of oolitic limestones and calcareous grits from the Corallian succession that has outcrops nearby.***

LEFT ***St Gregory's Minster was built with Corallian limestones quarried nearby; parts date back to 1060, including the Saxon sundial over its main door.***

Rievaulx Abbey

Once the heart of an important Cistercian monastery, and now a magnificent ruin, Rievaulx Abbey stands on the wide floor of Ryedale, sheltered deep within the Hambleton Hills. It is a textbook case of site selection and construction defined by local geology at a time before good transport allowed hauling resources from afar. The abbey stands on a low terrace of Osgodby sandstone, which locally replaces part of the Oxford Clay; so the site is well drained, stands above the floodplain, and has springs from the base of the Corallian beds low in the slopes behind the adjacent village.

After their founding in 1132, the building of the various parts of the monastery and abbey was spread over most of a hundred years. Early construction used Hackness Rock, part of the Osgodby sandstone, that was extracted from a quarry near Bow Bridge. By 1145 this was being sent the 500 metres to the abbey in barges on a canal built for the purpose. The stone could be dressed into good facing blocks, while the cores of the thick walls were filled with a rubble of hard Hambleton Oolite; this was taken from the wooded slopes directly above the abbey, but stone from these outcrops proved unsuitable for dressing.

The site of Rievaulx Abbey, with the quarries that supplied most of the stone for its construction, and the short-lived canals that provided links to the abbey.

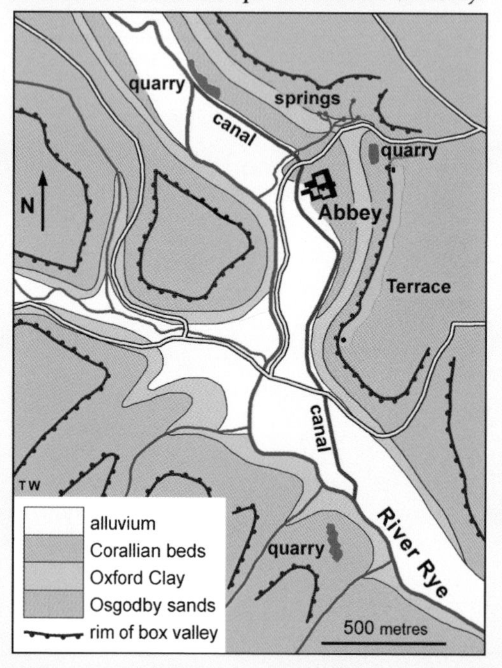

By 1150 a second phase of building used strong sandstone from the Saltwick sequence: this was quarried in Bilsdale, then carted downhill for nearly ten kilometres to the abbey. It was worth transporting this stone a longer distance because it was more easily dressed and carved. By about 1200 an even better stone, the Birdsall Grit, was being used. This was quarried south of the abbey, in an outcrop high on the hill of Hollins Wood, from where is was sledged down a steep track to the valley floor, then taken by barge on a second canal, a kilometre long, up the dale to the abbey.

With their Cistercian ethic of self-sufficiency, the monks made their own tools with iron from their own smelters. Their ore was a rather lean ironstone within the Eller Beck beds that crop out in lower Bilsdale. However, all came to an end with the Dissolution of the Monasteries in 1538. The abbey roof was taken off and the building left to crumble, while the monks were dispersed. Iron smelting did continue at Rievaulx until the dale's woodlands were stripped, leaving the smelters bereft of fuel – so they were closed down before 1650.

By the 1750s Rievaulx had become part of the Duncombe estate, when the 700-metre-long Terrace was carved out of the hillside 70 metres above the abbey. This swathe of lawn, with imitation temples at each end, was designed to offer grand views along Ryedale with the ruins of the abbey in the foreground below; sadly it has now been rendered almost pointless, as most of the views are obscured by large trees that have been left to grow in furtherance of bio-conservation.

ABOVE ***A long surviving face in the extensive old Spaunton Quarry that worked the thin unit of Coral Rag on the dip slope of the Tabular Hills.***

RIGHT ***One of the most photogenic houses in Yorkshire, Beck Isle Cottage was built with the local oolitic limestone to stand beside the stream in the village of Thornton Dale.***

BELOW ***The magnificent ruins of Rievaulx Abbey, built with stone that was largely derived locally within Rye Dale.***

fame when it broke into the bone deposits inside Kirkdale Cave. This was typical of olden times, when the choice of stone was dictated by the distance from the quarry. As transport improved, quarries in the better stone could thrive, and numerous small quarries were abandoned. East of Kirkbymoorside, Spaunton Quarry extended for over a kilometre until it ran out of planning permission and closed in 2007. It extracted Coral Rag, the hardest of the Corallian limestones, and also some of the overlying Upper Calcareous Grit.

Though once worked for dimension stone, production at Spaunton changed to aggregate, which is always in demand, while natural stone is little used in modern construction. The Coral Rag is strong enough for use in concrete, and extraction has continued at the Newbridge Quarry, 8 km east of Spaunton. Similarly, two quarries in the Vale of Pickering, one south of Wykekan and the other east of West Heslerton, continue to produce sand and gravel from the deltaic sediments accumulated on the floor of the Quaternary lake.

The lovely row of houses that lines the main road east out of the village of Thornton Dale was built with locally sourced Malton Oolite.

Stone in the Howardian Hills

Folding and faulting of the Jurassic rocks has produced a jumbled pattern of outcrops within the Howardian Hills, thereby providing considerable variety among building stones taken from numerous quarries. Standing in the heart of the Hills, the grand house known as Castle Howard is built of Birdsall Grit; this has long been the best of the calcareous grits within the Corallian sequence for the production of clean ashlar blocks. Although some beds were found not to weather so well, Birdsall Grit has been quarried at numerous sites and used extensively in the local buildings.

Malton Oolite was, of course, the key material for all the old houses in its eponymous town. Much of the stone came from the old Brows Quarry, close to the River Derwent on the western edge of town. The rather splendid colonnaded Town Hall in Malton was built with this oolite, though its steps have since been replaced with more durable Millstone Grit sandstone from the Pennines. Brows Quarry also worked the Lower Calcareous Grit, which lies directly beneath the oolite where the intervening sequence is locally absent. The quarry had a tramway down to a wharf on the Derwent, so was able to send stone some 25 km down the river for the rebuilding of Stamford Bridge in 1727. Though it ceased activity in

The very fine Town Hall in Malton, largely built with the Malton Oolite.

1948, and now a dense woodland is growing there, there has been a call to protect the quarry from any redevelopment because it could be a valuable resource for stone in restoration work on so many buildings in the region.

Further west, the Malton Oolite is again conspicuous. The attractive village of Slingsby has houses made of the warm-toned limestone, many dating from the early 1700s. Nearby, Hovinghan Quarry works the oolitic limestone and also its harder upper member, the Coral Rag, for aggregate for concrete, and also lime for cement; larger blocks are set aside as and when required to be taken off site to produce dimension stone. The adjacent Wath Quarry worked the same beds, but is no longer in use.

The Hildenley Limestone is a variant of the Malton Oolite, lying at its top just below the Coral Rag; it is only known in a small outcrop west of Malton. It is a homogeneous, fine-grained, compact white limestone and is claimed to be the finest stone in Yorkshire for decorative sculpture. It has been quarried since Roman times, and can be found as detailing, quoins, window frames, coffins and more in nearby pre-Norman churches at Amotherby, Appleton-le-Street and Hovingham; it is also beautifully displayed in the one surviving arch among the ruins of Kirkham Priory. The quarries were last worked in 1933 and are now shrouded in trees within Hildenley Wood.

Though the Ravenscar beds are dominated by sandstones, the relatively minor inter-bedded limestones have yielded the better building stone from outcrops in the Howardian Hills. The Whitwell Oolite is a thin bed of white limestone among the Cloughton sandstones; it was quarried around the village of Whitwell-on-the-Hill to be used extensively in nearby villages and also in parts of the structure of Kirkham Priory. Higher in the succession, Brandsby Roadstone was the local name for a thinly bedded, siliceous limestone that occurs towards the western end of the Howardian Hills; it is hard enough to have been used on the roads, and splits easily into slabs for local buildings, and in some cases even for roofing.

The ruins of Byland Abbey, which was built with calcareous grit from the Corallian succession.

Birdsall Grit was used to build Castle Howard and also the stately archways over its approach roads.

There is no good local stone around West Lutton in the heart of the chalk Wolds; St Mary's Church was built with sandstone in from the North York Moors.

Stone in the Wolds

Most of the chalk in south-eastern England is too soft and friable to constitute a decent building stone, but the same rock in eastern Yorkshire has a stronger mineral cement. It is therefore hard enough to have been used for building many of the earlier houses in Wolds villages, especially in those around Flamborough and Bridlington. Its main quarries were near Boynton, west of Bridlington (though Boynton Hall was built of brick with detailing of Hildenley Limestone). Flints are not abundant within the northern chalk, so the Yorkshire Wolds lack the village houses built with flints embedded in liberal amounts of mortar, which are so distinctive in England's southern chalklands. Even the harder northern chalk was generally unsuitable for building anything larger than small houses; however, a rare exception is the old lighthouse at Flamborough, one of the oldest that survives in England. Built in 1674, it was only ever a beacon, without any light beyond a fire in a brazier on its roof. It was built with roughly shaped rubble chalk, though window surrounds are of sandstone, and later repairs and restorations have used dressed stone and lime-washed brick.

Further to the south, the chalk has been little used as building stone. In the past, many small quarries have been worked for lime, and the few large quarries still active are there to feed cement production. Village houses are dominated by brick. Larger buildings, including most of the churches, have had to import stone from afar. The splendid minster at Beverley was built with Permian Magnesian Limestone (now known as the Cadeby Formation), the same stone that was used to build York Minster, both of which were supplied from quarries near Tadcaster.

The walls of Kirkham Priory were faced with finely dressed blocks of Hildenley Stone, but after the Dissolution of the Monasteries most of these were plundered to be used elsewhere; this wall core of a mix of poorer limestone and sandstone was left standing, though the facing stones were not all stripped away from the arch in the background.

The old lighthouse at Flamborough, one of the few tall structures to have been built with blocks of chalk.

With no good local stone on the Wolds, St Mary's Church in South Dalton was built around 1860 with Permian dolomitic limestone from a quarry near Worksop; its slender spire is 63 metres tall.

Building stone in Holderness

The simple statement is that there is no building stone in Holderness. This great plain of glacial debris, mud, peat and wetland has no hard material, save the scatter of boulders within glacial till and a few patches of gravel with pebbles of hard quartz. Holderness villages have their own character, but it is not one that relates to the local geology. Builders of the early churches required something more durable than timber, and some resorted to any available rounded boulders; these could be used in walls retained by quoins of dressed sandstone or limestone that was hauled in from afar. Some boulders were from the beaches, having been washed out of the glacial till; others were field stone, picked out of the till to leave a soil consisting of only finer materials; and some were garnered from abandoned ships' ballast.

Along the coast of Holderness it can be said that modern housing has been dictated by the area's geology in that the most appropriate are wooden bungalows or caravans. The limited lifespan of most wooden bungalows means that they are no great loss when undermined by erosion along the rapidly retreating coastlines, though some have been successfully dragged back from imminent doom. Even better are holiday caravans that can benefit from seafront locations and can simply be rolled back each year when winter storms remove yet more of their seafront: yet another way in which geological processes influence mankind's interaction with the natural landscapes.

A spectacular variety of hard stones, sedimentary, igneous and metamorphic, were picked out of local glacial till around Aldbrough, and used to build the walls of the Church of St Bartholomew, above lower courses of dressed sandstone.

CHAPTER 11

Landscapes to Enjoy

Moors and Wolds form the high ground, so are the natural targets for lovers of the great outdoors. But they are only highlights within the varied landscape of eastern Yorkshire. Of the two, the Moors are the senior partner: their great spread of undeveloped wilderness merits their National Park status. Visitor numbers to England's National Parks put the North York Moors in the middle rank; it lacks the big mountains, but does have a glorious coastline. The park is within easy reach of huge numbers of city dwellers, and it has one of Britain's greatest traditional seaside resorts at Scarborough. The Moors also benefit from adjacent areas with great landscapes, and those include the Yorkshire Wolds.

None of eastern Yorkshire's varied terrains is rugged enough to make travel seriously difficult, and farming has long supported a plethora of villages. Tiny hamlets and isolated farms lie along the valleys that dissect the high moors. All are now linked by roads that make the landscapes easily accessible, though an east–west journey across the North York Moors does trace a zig-zag line with some rather long loops.

Heather-covered uplands are the defining element of the Moors. At their eastern end, the expanse known as Fylingdales Moor rose to fame in 1962, when it acquired the Early Warning Radar Station, with its three huge spherical housings for the antennae. Designed to detect missiles inbound from Russia, they could give a few hours' warning to targets in America, and incidentally give four minutes' warning to targets in Britain. This gave rise to many jokes about what to do in those four minutes, best of which was the line in the review show, *Beyond the Fringe*, that 'some people can run a mile in four minutes' – this was not long after Roger Bannister's great performance at Oxford. Visible from afar by travellers on the main road from Pickering to Whitby, the giant golf balls could have been welcomed as a rather glorious piece of landscape art, were it not for their military infamy. Sadly they were demolished in 1992, and replaced by a huge new radar housing that is distinctly unlovely.

The 'great outdoors' is one of the key attractions of eastern Yorkshire. Whereas the Moors attract the more hardy walkers, the entire coastline, alongside both Moors and

Moorland by-road, heading north from Hutton-le-Hole across Spaunton Moor towards Rosedale.

The golf balls of Fylingdales Moor, demolished in 1992.

Wolds alike, offers a huge variety of shorter strolls, when the specialists have their own venues. Rock climbers head to grit crags around the outer edge of the Cleveland Hills, where Highcliff Nab, above Guisborough, is one of those that benefited from bygone quarrymen creating high walls of clean rock. Cavers probe modest attractions beneath the Tabular Hills, kayakers find white water on parts of the River Esk, and gliders launch their gliders into the wind from the crest of Sutton Bank. Meanwhile fossil collectors comb the foreshore and cliffs along England's northern version of a Jurassic Coast.

A surprising number of local bus services across the Moors and along its coastline make it possible for walkers to enjoy grand treks through the countryside without having to loop back to a remote car park. A novel alternative to car or bus is the North Yorkshire Moors Railway, linking Pickering to Grosmont and now through to Whitby. It is a significant visitor attraction, though not all its steam-engine enthusiasts appreciate that the trains puff up and down one of England's finest Ice Age landforms: Newtondale is the perfect example of a channel cut by glacial meltwater.

Lower Calcareous Grit offers strenuous climbing on Park Scar, in the side of a valley tributary to Rye Dale.

Conservation and protection

The North York Moors were designated as a National Park in 1952, extending over the Cleveland Hills, the Hambleton Hills and the Tabular Hills. Unlike the preserved wilderness in some foreign lands, a National Park in Britain is a working landscape with its villages and farms intact, though industry is kept to a minimum. Planning and development rests in the hands of the Park Authority, hence apart from any dominance by distant cities. So the North York Moors retain their rural character, their lovely landscapes and their cultural heritage, while also recognising the needs of visitors, and incidentally benefiting from the economy of tourism on an appropriate scale.

Into the North York Moors National Park with the White Horse of Kilburn high on the sandstone hill.

The next level down in environmental protection is designation as an AONB (Area of Outstanding Natural Beauty). Like National Parks, these are lauded for their landscape values and the environment is protected, but their management does not include planning powers and they have less scope for outdoor recreation. The Howardian Hills were given AONB status in 1987, since when they have retained their rural character while seeing relatively few tourist visitors other than to the stately home of Castle Howard. Designation as an AONB is currently in the planning stage for a large part of the Yorkshire Wolds, which will include the high ground that is dissected by all the deepest of the beautiful dry valleys.

A local bus crosses Danby High Moor on the service route from Rosedale to Eskdale.

A Heritage Coast has no statutory powers, but is recognised as an area of significant natural beauty and draws attention to its needs in environmental protection. The Flamborough Headland Heritage Coast justifies its nomination with its splendid length of chalk cliffs. These contrast with the rather featureless hinterland that is the eastern end of the Yorkshire Wolds, where the chalk is largely shrouded with glacial till, and will probably not be part of the planned AONB.

There are nature reserves scattered across the Moors, Wolds and intervening areas. The only National Nature Reserve in the region is Spurn, covering the entire, long, thin peninsula out to Spurn Head. Others are cared for by various organisations. Notable is the coastal site of Bempton Cliffs, between Filey and Flamborough, justifiably famous for its seabirds. Black guillemots, small grey kittiwakes and larger yellow-headed gannets

ABOVE Gannets galore build their precarious nests on tiny, narrow ledges offered by the bedding structure in the towering chalk cliffs at Bempton.

LEFT Large modern farms stand below the unproductive moorland that surrounds the head of Little Fryup Dale.

dominate; most build nests of moss and plant debris, but guillemots just place their eggs on the bare rock. And a few puffins find homes in narrow fissures, or dig their burrows into the glacial till above. The site is managed by the Royal Society for the Protection of Birds, with a visitor centre and viewing platforms that overlook the spectacular cliffs.

The North York Moors have long been famous for their heather, which creates swathes of brilliant purple in late August and early September, before returning to its winter colour of sombre brown. Sadly, the heather purple is not what it used to be. A warmer, drier climate does not suit the heather, and since 2019 it has also fallen prey to heather beetles. Winters have not been cold enough to kill off the beetles' larvae, which devour the heather leaves, stress the plants and reduce the vibrancy of their flowering. The heather might recover from its beetle attack, but is also threatened by the controversy over management versus wilding.

Young heather provides nesting and feeding for red grouse and various other moorland birds, so it is sad that the moors have become a popular venue for shooting the game birds. However, the economy of organised shooting parties funds management of the moors. This entails ensuring the bloom of new heather by clearing the old growth that would otherwise mask the new. Traditionally this has been done by controlled burning, which is opposed by some, on the grounds of its contribution to global warming. The alternative to burning is swiping or flailing with blades or chains; this eliminates production of carbon dioxide, but is not practicable on steeper ground. Without management of the heather, or even without the grouse shooting to fund the management, the moors will revert to a scrubland, and ultimately to woodland of willow, birch and other hardy species. An uncontrolled 'return to nature', within the current fervour for 're-wilding', would destroy a landscape where mankind has actually made a positive contribution. It remains to be seen whether or not the heather moors have a long-term future.

Heather displays its August purple on Westerdale Moor.

Walk the landscape

A huge network of footpaths allows walkers to really appreciate the landscapes. The North York Moors and the Cleveland Hills also have great tracts of access land, with freedom to roam across the high plateaux; only the valley floors are farmed, though are still crossed by specified paths. The converse applies in the Wolds, where the access land is restricted to the deeper and steeper valleys, but these provide the best walking anyway. And there is almost no access land on the hills between Moors and Wolds, where farmland cover is complete except in the woodlands of Dalby Forest.

Long-distance footpaths offer sign-posted routes that take in many of the best features. Of these designated routes, the Lyke Wake Walk is perhaps the most famous, linking markers near Osmotherley and Ravenscar by following the Moors watershed for some 64 km. Initiated in 1955, this route is one for the hardy walkers, as it is designed to be a challenge to be walked within 24 hours. It is not sign-posted, though its route is well trodden and traceable, but it is not really a walk for enjoying the landscape and geology in the rush to complete the distance.

Almost the opposite is the Coast-to-Coast Walk, devised by that doyen of ramblers, Alfred Wainwright, to traverse the three

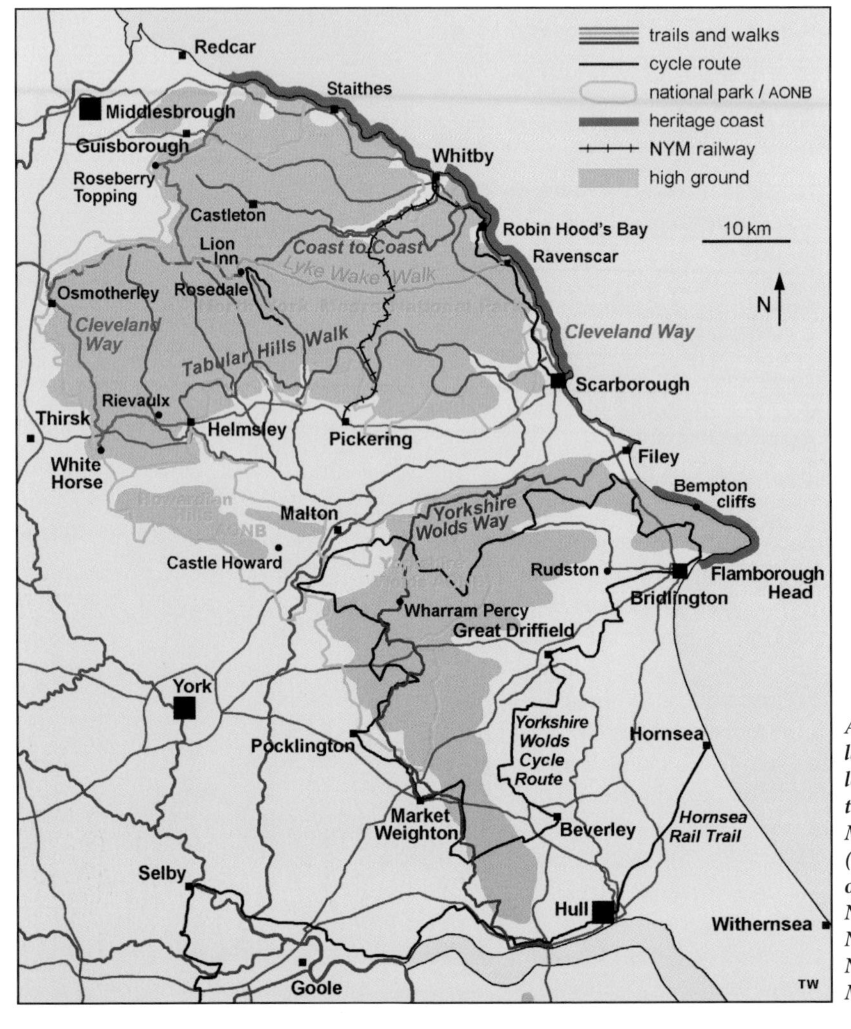

Areas of protected landscapes and long-distance trails within the Moors and Wolds. (AONB = Area of Outstanding Natural Beauty; NYM railway = North Yorkshire Moors Railway.)

A signpost for footpaths where the Yorkshire Wolds Way climbs out of the chalk valley of Thixendale.

great National Parks of northern England: the Lakes, the Dales and the Moors. Though unofficial, it has proved to be so popular that it is due to receive National Trail status, and will be upgraded prior to a formal opening in 2025. After a trudge across the Vale of York, it climbs on to the Moors at Osmotherley along with the Lyke Wake Walk and the Cleveland Way, to follow the high ground that offers wide vistas to the north. It then joins the old ironstone railway to Rosedale, near the Lion Inn.

The walk down the long ridge to Glaisdale offers scope for visualising the great Ice Age lake that occupied Eskdale, before a stroll down the dale itself. Then return to the high ground to Littlebeck, and a worthwhile diversion via The Hermitage, a cave carved into a bluff of Dogger sandstone, and the Falling Foss waterfall. Finally the route takes you back on to an arm of the moors, and onwards to a glorious finish at Robin Hood's Bay.

Part of the Coast-to-Coast follows the Esk Valley Walk, which heads up the dale from Whitby to Casteleton, before making a loop along the ridges round both sides of Westerdale, via the obligatory break at the Lion Inn.

A host of other walks, long or short, have maps and guides readily available on-line. Notable are the region's three senior walks along the crests of the great escarpments that dominate the landscapes of eastern Yorkshire. These long-distance trails are defined by the geology, but they do require a commitment to major treks, whereas some short sections of coastal footpaths offer more to see of the geology and geomorphology.

Stepping stones on the River Esk; an option during dry-weather for walkers on the Eskdale Way; the footbridge can be welcome when the river runs high.

The Cleveland Way, following the ancient trackway of Hambleton Street along the crest of the Hambleton Hills.

Cleveland Way

This National Trail extends for 176 km in a great loop along the crest of the Cleveland Hills, and then beside the coast that fringes the North York Moors. It starts at Helmsley with a traverse of the Hambleton Hills, where a short spur takes in the White Horse above Kilburn. This giant hill carving was made in 1857 as a northern copy of the chalk figures in southern England; however, it is not quite genuine because it is on a sandstone slope, so required a coating of chalk chippings after the turf was cut away.

The trail then heads north on the high ground of Corallian grits and limestones, before a descent into Osmotherley. This is followed by a rise on to the crest of the Cleveland Hills formed by the main escarpment of Ravenscar sandstones, culminating at the ragged outcrop that has been weathered and broken to form the dramatic chaos of the Wainstones. Northbound again and then across Kildale, the way-marked trail heads for a short spur up to Roseberry Topping. That splendid landmark is formed by a cap of Ravenscar sandstone atop a cone of Whitby Mudstone that was steepened by its great landslide in 1912.

Descending through Guisborough Woods, the trail leaves the sandstone heights of the Cleveland Hills to enter a rolling landscape that was a hive of industry 150 years ago. This was the heartland of mining the Cleveland Ironstone to feed the steelworks of Teesside, though today there is little trace of the bygone mines.

At Saltburn, the Cleveland Way meets the coast, which it then follows southwards to a grand terminus on the end of Filey Brigg. Largely along cliff-tops, the trail is endlessly diverse. Headlands of strong sandstone alternate with bays cut into weak, fossiliferous mudstones, and more than 150 landslides are listed along this one short section of coast. Many are formed where weak rocks have sheared beneath strong ones, but others are simple rotational failures within the weaker mudstones. Some are still active, but most have stabilised and have blended into the landscape.

Those natural scars are interspersed with man-made scars where alum shales were quarried on a huge scale at Loftus, Boulby, Kettleness, Ravenscar and elsewhere. Most of these are also now blending into the landscape, and are only apparent to walkers who are wondering what this coastal landscape looked like perhaps 200 years ago.

Walkers on the Yorkshire Wolds Way along one of the lovely dry valleys that are tributary to Millington Dale.

Yorkshire Wolds Way

Extending for 127 km, the Yorkshire Wolds Way offers a gentler style of walking, simply because chalk downlands are less rugged than sandstone moors. The trail starts beside the River Humber, right beneath the soaring suspension bridge; between 1981 and 1998 this bridge was the longest in the world with its span of 1410 metres. Northwards, the trail loops on and off the Wolds before skirting Market Weighton and passing the Rifle Butts Quarry, with its classic exposure of Red Chalk.

A low-level traverse past Pocklington is followed by a rather circuitous route that takes in many of the finest of the chalk dry valleys between Millington and Thixendale. Spiral earth ridges on the floor of Thixendale are neither natural nor ancient, as they are an item of landscape art created in 2011. Reality is regained a little further north where the trail passes the excavated remains of the abandoned medieval village of Wharram Percy.

The last third of the trail lies largely along the crest of the chalk escarpment, with grand views north across the Vale of Pickering. An abundance of tumuli and linear earthworks appear to date from Iron Age settlement and farming that was successful on the chalk. A final descent from the Wolds leads down to Muston and a trail finish at Filey.

Tabular Hills Walk

Between the Moors and the Wolds, the Tabular Hills have their own designated trail that largely traverses the high ground of the Corallian grits and limestones. Some 77 km of way-marked footpaths link Helmsley through to Scarborough. The western section involves plenty of hill climbing where it cuts across the ridges and valleys. This makes very clear the extent to which rivers draining south from the Moors have dissected the low plateau of the Tabular Hills, which is actually the very gentle, south-facing dip slope of the Moors escarpment.

After crossing the magnificent Newtondale meltwater channel, the Walk rises to a high point on Levisham Moor, with grand views northwards from the crest of the escarpment, though the nearest section of the Moors are shrouded in forest. Better views are to the south, where the trail traces the rim of the enigmatic Hole of Horcum.

Further east the trail runs for great distances through forests and plantations, until it descends into the Upper Derwent valley, just before the natural course of the river turns south into the Ice Age meltwater channel of Forge Valley. Avoiding this, the Walk follows the Sea Cut, the drainage channel created to divert floodwater away from the Vale of Pickering; both channel and trail end on the coast at the northern edge of Scarborough.

'Waves and Time', the landscape artwork beside the Yorkshire Wolds Way on the floor of Thixendale.

The softer face of the North York Moors along the coast, with Staithes far left and Boulby mine on the right.

Shorter coastal walks

The English Coast Path is a grand project still in progress, which will eventually establish a footpath along the entire coast; however, at present only dissociated sections are designated. A novel feature is that it will be a rolling path, in that its right of way will remain intact should it have to move inland wherever coast erosion causes cliff retreat. This will be particularly valuable along the Holderness coast, where chunks of the coastal path fall into the sea every winter. Many sections of this active coastline are accessible, and the short walk from Skipsea to Ulrome has for some years been among the finest to appreciate coastal erosion in progress.

Routes between Robin Hood's Bay and Ravenscar.

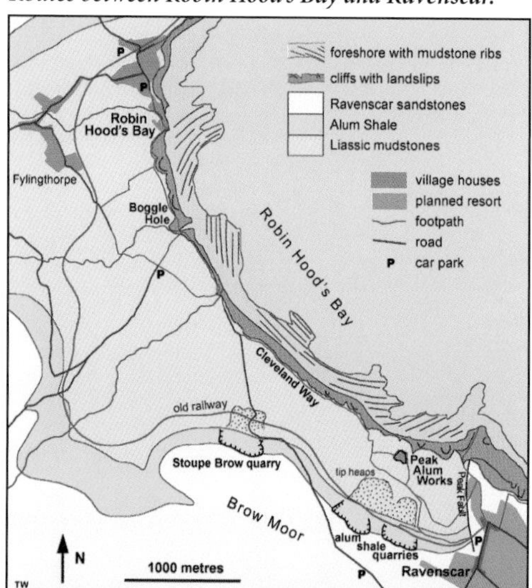

There is endless scope for walking the Yorkshire coast, but three half-day walks can be recommended for sheer enjoyment and appreciation of the diverse geology.

Robin Hood's Bay offers an excellent circular walk of about twelve kilometres via Ravenscar. From the car park at the top end of the new part of the village of Robin Hood's Bay, head south along the old railway line, with sweeping views from the high level. Pass the old quarries in Alum Shale at Stoupe Brow, though they are now almost lost in woodland. Further south, the two large quarries beside the trail yielded the alum shale to supply the Peak Alum Works for most of 250 years, until activity finally ended in 1862. Continue to the tiny village of Ravenscar, where a Victorian resort town was planned but never happened. Then head back north on the Cleveland Way footpath, which is signposted to pass beside the partly restored remains of the old alum works. The path along the cliff tops looks down on to the foreshore, with its ribs of mudstone curving round where they cross the Cleveland Anticline.

The Cleveland Way northbound to Robin Hood's Bay.

Queen's Rock, a tall sea stack of chalk within a small bay east of North Landing on Flamborough Head.

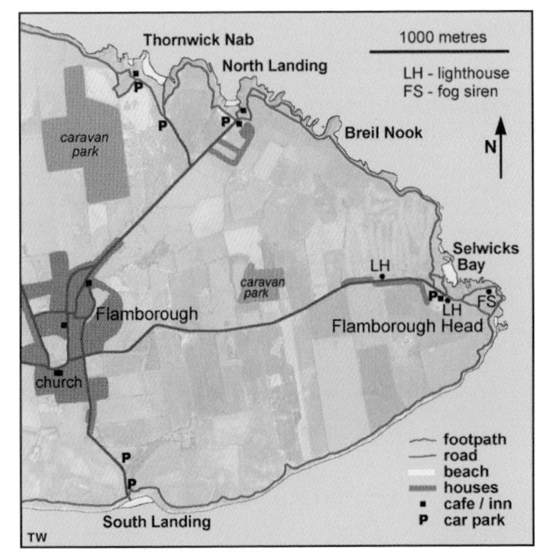

ABOVE ***Cliff-top footpaths around Flamborough Head.***

A low-tide option north from Boggle Hole is to walk along the beach, with opportunities to search for fossils in the Redcar Mudstone, though the cliffs are mainly of slumped glacial till. Then end with a steady walk up the narrow street of Robin Hood's Bay's old village, past many shops with fossils on display and for sale.

Flamborough Head can be enjoyed in an almost level, circular walk of twelve kilometres, with access points at car parks at each road's end. From Thornwick Nab right round to South Landing the cliff-top path offers easy access and great views, with the best of the headlands, bays, inlets, geos, stacks

BELOW ***Thornwick Nab on the walk round Flamborough.***

and arches strung out all along the northern sector. At low tide, a descent to the foreshore in Selwicks Bay is worth the effort of the return. A wide, wave-cut platform is scarred by ribs of gently folded chalk, and provides access to the marine arches in the northern headland. Close to the steps down to the beach, the cliffs expose fault lines and calcite veins on a scale rarely seen within the chalk.

Spurn Head is no longer accessible by car, since the road was partly eroded and partly buried during a major storm event in 2013. It is a full six kilometres from the road end out to the tip of the headland, just beyond the lifeboat station that was closed in 2023. The entirity of this magnificent sand spit can only be appreciated on a long walk in a remote setting. But for those who enjoy sand, wind, big skies and a few seabirds, Spurn Head offers a remote experience unlike any other in Britain.

Cycle the landscape

A great opportunity for traffic-free cycle trails came with re-surfacing abandoned railway lines, to give easy access on gentle gradients to great cross-sections of the terrain. Some of the trails have become popular, and there are cycle-hire outfits at some key locations.

The old line from Scarborough to Whitby winds around high hillsides, offering panoramic views of various parts of the coastline; it passes right by the quarries in alum shale at Ravenscar, and ends with a ride over the Larpool Viaduct high above the River Esk.

Very different is the Hornsea Rail Trail. This takes an almost straight line across Holderness, and is the best way to see the expanse of lowland where distances are too great for interesting walking. The trail links Hornsea to Hull, but extends to Selby and beyond as the Trans Pennine Trail.

Old and older lighthouses near the end of the long sand spit out to Spurn Head.

Cycle track along the old ironstone railway that loops round the head of Rosedale.

The old Rosedale mineral railway is not a designated trail, but is well surfaced and is easily accessible where it crosses the heights of Blakey Moor. Eastwards, the two branches contour high along both sides of Rosedale, and pass beside the spectacular remains of all three sets of old kilns that were built to process the ores (*see* map on page 116). The ends of the two branches are linked by minor roads passing through Rosedale Abbey (the village that has never had an abbey) to make a splendid loop of 18 km, though some effort is required on the famously steep road up Chimney Bank, rising to the end of the western railway at Scar Top.

Westwards from Blakey Moor, cyclists can join walkers on the Lyke Wake or the Coast-to-Coast to follow the railway trail across the high and remote moors round the top end of Farndale. But then, after the walkers' trails diverge, the route north forgoes gentle gradients where it descends 220 metres off the edge of the Cleveland Hills on what was originally an incline with a winding house and rope-haulage of the wagon trains.

Beyond the old railway track-beds, the National Cycle Network is being developed as sign-posted routes that keep to back-roads with minimal traffic. The Yorkshire Wolds Cycle Route makes a loop of 230 km, out of and back to Beverley, entirely on minor roads. The best of the high Wolds are traversed, and include the lovely road that winds the whole length of Middleton Dale, finest of the dry valleys. It takes a loop out to Kirkham Abbey, set in its great meltwater channel, and offers a break from the saddle at the bird cliffs of Bempton. On the return leg across the lower flanks of the Wolds, an optional route for the geo-minded cyclist would be to go a few extra kilometres to take in the great monolith at Rudston, and ponder its glacial origins.

Whether in the saddle or on foot, or even falling back on to car or bus, the Moors and Wolds of eastern Yorkshire offer a wealth of glorious landscapes. Features of Ice Age erosion vie with relics of bygone mining. Then, behind all the detail, the wider features of contrasting terrains are defined by the underlying geology. And that is what landscape is all about.

End of the road at Aldbrough – quite literally so, with continuation of the tarmac lost to the eroding coastline of Holderness.

Further Reading

The basic resource text on the geology of the Moors and Wolds is the *British Regional Geology of Eastern England, from the Tees to the Wash*, published by the British Geological Survey. The current edition was published in 1980, so some of the stratigraphic names have since changed. However, updated geological maps at 1:50,000 cover the whole area, and are now free to view at webapps.bgs.ac.uk/data/maps.

The book by members of the Yorkshire Geological Society (*The Geology and Mineral Resources of Yorkshire*; edited by D. H. Rayner and J. E. Hemingway) was published in 1974, so some of its stratigraphic names are also dated, but it is a valuable and very comprehensive resource (at 405 pages); long out of print, second-hand copies are worth looking for. Published by the same Society in 1994, *Yorkshire Rocks and Landscape*, edited by Colin Sutton, is a field guide that includes detailed descriptions of walks along six sections of the coast between Staithes and Flamborough.

An authoritative overview of the geomorphology is Allan Straw's section on Eastern England in *Eastern and Central England*, by Straw and Clayton, published by Methuen in 1979. More recent data on the Vale of Pickering is the 180-page Field Guide, *The Quaternary of the Vale of Pickering*, written by P. Lincoln, L. Eddey, I. Matthews, A. Palmer and M. Bateman, and published by the Quaternary Research Association in 2017. Plans and means of stabilising the Holderness coast are described at eastriding.gov.uk/coastalexplorer/pdf/1.

Late August on the heather moors above Westerdale, upper Eskdale

The wave-cut platform exposed at low tide in Selwicks Bay, Flamborough.

There are numerous books and on-line items on the various mining industries in and around the North York Moors, of which the following are reliable and useful:

Catalogue of Cleveland Ironstone Mines, with 80 pages written and published by Peter Tuffs in 2005.

The Alum Industry of North-East Yorkshire, by Peter Appleton in 2018, at east-clevelands-industrial-heartland.co.uk.

Geological background to the North Yorkshire alum industry, by Denis Goldring, in *The Cleveland Industrial Archaeologist*, 2012, Number 33, pages 43–65.

Mining in the Zechstein evaporites at Boulby Mine, by Neil Rowley, *Proceedings of the Open University Geological Society*, 2016, Volume 2, pages 55–61.

For information on the jet gemstone, avoid the numerous commercial sites that are sales outlets, and refer to Sarah Steele's excellent website at eborjetworks.co.uk.

Building stones used in the Moors and Wolds are described in *A Building Stone Atlas of North-east Yorkshire*, written by John Powell in 2012, and available as a free download at nora.nerc.ac.uk/id/eprint/504439. Revision in 2023 yielded free on-line reports covering the Moors in *North Yorkshire, East* and the Wolds in *East Yorkshire and Northern Lincolnshire*, both available at historicengland.org.uk.

For a delightful read about parts of the Moors that you did not know, obtain a copy of *North York Moors above and below*, written by John Dale and Peter Ryder in 2022 and available from broomlee.org.

For topographic maps, the North York Moors are conveniently covered by two sheets, OL26 and OL27 in the Ordnance Survey Explorer 1:25,000 series; however, full coverage of the area to the south requires multiple sheets at any scale.

Then go out to the Moors, the Wolds, the hills and their coasts to enjoy the real thing.

Index

Entries in *italics* are photographs

First published in 2024 by
The Crowood Press Ltd
Ramsbury, Marlborough
Wiltshire SN8 2HR

enquiries@crowood.com

www.crowood.com

British Library Cataloguing-in-Publication Data

A catalogue record for this book is available from the British Library.

ISBN 978 0 7198 4374 7

Front cover: On the northern edge of the Cleveland Hills, Roseberry Topping is an isolated crag scarred by a history of mining and landslides.

Frontispiece: Purple heather crowds into the Cleveland Way footpath towards Skinningrove along the northern coast of the North York Moors.

Contents page: Gannets nesting on the chalk cliffs at Bempton, on the edge of the Yorkshire Wolds.

Typeset by Tony Waltham
Cover design by Blue Sunflower Creative
Printed in India by Parksons Graphics Pvt Ltd

Acknowledgements

This book is dedicated to the memory of Jan, my wonderful wife for 46 years; her boundless support and her many contributions made the book possible, but she died shortly before its pages were completed.

Grateful thanks are recorded to John Dale for his extensive support and guidance, both based on his huge knowledge of the North York Moors, and to Anthony Raithby for allowing generous use of his aerial photographs at arfotolog.smugmug.com.

Image credits are to John Dale on pages 41, 55 (all 3), 56B, 63T, 113, 114B, 126L, 127T and 129; Anthony Raithby on pages 6, 34, 39T, 48B, 73, 87 and 103B, Peter Ryder on pages 15L, 16tr, 17TL and 20B; Natural History Museum on page 20T; Great Ayton Historical Society on page 42T; and Komatsu Joy on page 122B. All other photographs are from the author's Geophotos collection.

The maps in these pages have been abstracted and compiled from numerous sources, before all were redrawn and embellished by the author. Credit belongs to members of the British Geological Survey for their source information and mapping.